VIVO

A Semantic Approach to Scholarly Networking and Discovery

Synthesis Lectures on Semantic Web: Theory and Technology

Editors
James Hendler, *Rensselaer Polytechnic Institute*
Ying Ding, *Indiana University*

Synthesis Lectures on the Semantic Web: Theory and Application is edited by James Hendler of Rensselaer Polytechnic Institute. Whether you call it the Semantic Web, Linked Data, or Web 3.0, a new generation of Web technologies is offering major advances in the evolution of the World Wide Web. As the first generation of this technology transitions out of the laboratory, new research is exploring how the growing Web of Data will change our world. While topics such as ontology-building and logics remain vital, new areas such as the use of semantics in Web search, the linking and use of open data on the Web, and future applications that will be supported by these technologies are becoming important research areas in their own right. Whether they be scientists, engineers or practitioners, Web users increasingly need to understand not just the new technologies of the Semantic Web, but to understand the principles by which those technologies work, and the best practices for assembling systems that integrate the different languages, resources, and functionalities that will be important in keeping the Web the rapidly expanding, and constantly changing, information space that has changed our lives.

Topics to be included:

- Semantic Web Principles from linked-data to ontology design

- Key Semantic Web technologies and algorithms

- Semantic Search and language technologies

- The Emerging "Web of Data" and its use in industry, government and university applications

- Trust, Social networking and collaboration technologies for the Semantic Web

- The economics of Semantic Web application adoption and use

- Publishing and Science on the Semantic Web

- Semantic Web in health care and life sciences

VIVO: A Semantic Approach to Scholarly Networking and Discovery
Katy Börner, Michael Conlon, Jon Corson-Rikert, and Ying Ding
2012

Linked Data: Evolving the Web into a Global Data Space
Tom Heath and Christian Bizer
2011

VIVO: A Semantic Approach to Scholarly Networking and Discovery

Katy Börner, Michael Conlon, Jon Corson-Rikert, and Ying Ding

ISBN: 978-3-031-79434-6   paperback
ISBN: 978-3-031-79435-3   ebook

DOI 10.1007/978-3-031-79435-3

A Publication in the Springer series
SYNTHESIS LECTURES ON SEMANTIC WEB: THEORY AND TECHNOLOGY

Lecture #2
Series Editors: James Hendler, *Rensselaer Polytechnic Institute*
                Ying Ding, *Indiana University*
Synthesis Lectures on Semantic Web: Theory and Technology
ISSN pending.

VIVO

A Semantic Approach to Scholarly Networking and Discovery

Katy Börner
Indiana University

Michael Conlon
University of Florida

Jon Corson-Rikert
Cornell University

Ying Ding
Indiana University

SYNTHESIS LECTURES ON SEMANTIC WEB: THEORY AND TECHNOLOGY #2

ABSTRACT

The world of scholarship is changing rapidly. Increasing demands on scholars, the growing size and complexity of questions and problems to be addressed, and advances in sophistication of data collection, analysis, and presentation require new approaches to scholarship. A ubiquitous, open information infrastructure for scholarship, consisting of linked open data, open-source software tools, and a community committed to sustainability are emerging to meet the needs of scholars today.

This book provides an introduction to VIVO, http://vivoweb.org/, a tool for representing information about research and researchers—their scholarly works, research interests, and organizational relationships. VIVO provides an expressive ontology, tools for managing the ontology, and a platform for using the ontology to create and manage linked open data for scholarship and discovery. Begun as a project at Cornell and further developed by an NIH funded consortium, VIVO is now being established as an open-source project with community participation from around the world. By the end of 2012, over 20 countries and 50 organizations will provide information in VIVO format on more than one million researchers and research staff, including publications, research resources, events, funding, courses taught, and other scholarly activity.

The rapid growth of VIVO and of VIVO-compatible data sources speaks to the fundamental need to transform scholarship for the 21st century.

KEYWORDS

scholarship, research, research networking, Semantic Web, linked open data

Contents

3 Implementing VIVO and Filling It with Life 35

Valrie Davis, Kristi L. Holmes, Brian J. Lowe, Leslie McIntosh, Liz Tomich, and Alex Viggio

6 Extending VIVO 85

Chris Barnes, Stephen Williams, Vincent Sposato, Nicholas Skaggs, Narayan Raum, Jon Corson-Rikert, Brian Caruso, and Jim Blake

7 Analyzing and Visualizing VIVO Data

Chintan Tank, Micah Linnemeier, Chin Hua Kong, and Katy Börner

Dean B. Krafft, Katy Börner, Jon Corson-Rikert, and Kristi L. Holmes

Preface

The world of scholarship is changing rapidly. Increasing demands on scholars, the growing size and complexity of questions and problems to be addressed, and advances in sophistication of data collection, analysis, and presentation require new approaches to scholarship. A ubiquitous, open information infrastructure for scholarship, consisting of linked open data, open-source software tools, and a community committed to sustainability are emerging to meet the needs of scholars today.

VIVO provides an expressive ontology, tools for managing the ontology, and a platform for using the ontology to create and manage linked open data for scholarship and discovery.

Begun as a project in the Cornell University Library in 2003, VIVO is a tool for representing information about research and researchers—their scholarly works, research interests, and organizational relationships. In 2009, the National Institutes of Health awarded a two-year $12.2M American Recovery and Reinvestment Act Award to a consortium including Cornell University, Indiana University, the University of Florida, Weill Cornell Medical College, Washington University School of Medicine in St. Louis, The Scripps Research Institute, and Ponce School of Medicine. The consortium developed VIVO as open- source software with sophisticated visualization and data harvesting tools to provide a platform for next-generation scholarship. Following the NIH funding, VIVO is being established as an open-source project with community participation from around the world. We estimate that by the end of 2012, over 20 countries and 50 organizations will provide information in VIVO format on more than one million researchers and research staff, including publications, research resources, events, funding, courses taught, and other scholarly activity.

The rapid growth of VIVO and of VIVO-compatible data sources speaks to the fundamental need to transform scholarship for the 21st century. In this book, we provide an introduction to VIVO. We hope you will consider participating in the creation of a new and open information infrastructure for scholarship.

Katy Börner, Michael Conlon, Jon Corson-Rikert, and Ying Ding
August 2012

Structure of the Book

This book provides a detailed introduction to VIVO, a software system for addressing connections and discovery in scholarship.

Chapter 1—Scholarly Networking Needs and Desires

Scholarship involves identifying questions to be explored, assembling teams for the exploration, using previously collected information, and advancing knowledge. Gaps in current practice are identified along with a means to improve scholarship through the use of linked open data and VIVO.

Chapter 2—The VIVO Ontology

The ontology for representing scholarly activity using VIVO is described as well as the processes for extending that ontology.

Chapter 3—Implementing VIVO and Filling it with Life

Tools and processes for getting data from various sources into VIVO are presented. Keeping VIVO synchronizing with existing databases and systems provides scholars with up-to-date information for connections, sharing, and discovery.

Chapter 4—Case Study: VIVO at the University of Colorado

The University of Colorado has implemented VIVO on its own, outside of the VIVO project funding. They describe their work in both technical and policy terms as related to existing information management systems and faculty engagement.

Chapter 5—Case Study: VIVO at the Weill Cornell Medical College

VIVO has been implemented at this prominent medical school and multi-institution Clinical and Translation Science Center (CTSC). This chapter discusses the weaknesses of a legacy researcher profile system and how the introduction of VIVO aims to meet existing challenges.

Chapter 6—Extending VIVO

VIVO is an open-source software system written in the Java programming language and capable of delivering data to a variety of popular open-source web tools. Methods for expanding the capabilities of VIVO by extending its Java source code and re-purposing its data are presented.

Chapter 7—Analyzing and Visualizing VIVO Data

VIVO data can be consumed by specialized tools internal or external to VIVO. Techniques for reusing VIVO data in analysis and visualization are detailed, including the social networking visualizations available in VIVO.

Chapter 8—The Future of VIVO: Growing the Community

VIVO is a socio-technical infrastructure embraced and shaped by a living, breathing community across the world. How can an international scholarly information architecture be adapted and sustained to meet the continuously changing needs of scholars? Where are we headed?

Acknowledgments

VIVO is the work of a talented and committed team. At the center is Jon Corson-Rikert, whose vision and quiet determination created the original concept and pilot implementation at Cornell. He has provided an intellectual and moral core to all the work that followed. Elaine Collier at the NIH saw the need for research networking and the opportunity afforded the scholarly community by the American Recovery and Reinvestment Act. Her leadership in developing the Request for Applications led to the support that enabled VIVO to move beyond its humble beginnings. Once funded (NIH/NCRR Award UL24 RR029822), the NIH project team of over 120 leaders, librarians, ontologists, developers, writers, outreach specialists, and managers has been a marvel of ingenuity, insight, and effort. Their work has resulted in a world community of VIVO adopters, developers, and thought leaders. Connected scholarship enabled by the Semantic Web through VIVO is being realized as a result of these pioneers. Special thanks go to Todd Theriault for his careful proofreading and to Samuel T. Mills for figure formatting. The text has been much improved by their significant efforts.

Acknowledgements

CHAPTER 1

Scholarly Networking Needs and Desires

Michael Conlon, *University of Florida*

Abstract

We live in an increasingly connected world. From the beginning of science, through the birth of journals and scholarly societies, to the current time, scholars have increasingly relied on connections among themselves, their ideas, and their works. The Internet provided a new foundation for scholarship as a standard for physical transport and connectivity, but in its first 20 years did not have standards for information representation and data sharing. The Semantic Web provides this next level of the information exchange standards stack. VIVO capitalizes on this new capability to provide an information layer for scholarship and the connections of people and ideas across the world.

Keywords

scholarship, research, research networking, Semantic Web, linked open data.

1.1 THE WORLD OF THE SCHOLAR TODAY

The Digital Academy [1] is a reality. Scholars are connected electronically to colleagues, students and mentors, funding agencies, and administrators. Time is of the essence. Tenure clocks are ticking, scholars need to publish or perish, competitors are responding to the very same funding solicitations, while conferences and workshops daily request participation and presentations. New results are appearing at increasing rates. Assessment is everywhere—periodic reports to provosts, department chairs, research groups, and funding agencies are the norm. Impact factors play an increasingly important role in evaluation. Information complexity and linkage density are increasing. The amount of material to be mastered for expertise in a discipline is growing, as is the specialization of disciplines. The Internet has provided extraordinary opportunities for scholarship—reducing barriers of time and distance and creating new opportunities for access to information and discovery. With those opportunities come expectations of increased scholarly productivity.

Information for scholars—and about scholarly activity—has not kept pace with the increasing demands and expectations. Information remains siloed in legacy systems and behind various access controls that must be licensed or otherwise negotiated before access. Information representation is in its infancy. The raw material of scholarship—the data and information regarding previous work—is not available in common formats with common semantics.

The word processor and other tools of scholarship were designed to replicate a paper-based world. Web-based tools are very new. Although they capitalize on the ability of the Internet to remove time and distance constraints, they often require presence at specific sites and localized identity management, while ignoring the needs of scholars to access, replicate, and manipulate the previous work of other scholars.

Finally, scholars must repeatedly perform menial tasks to satisfy administrative requirements. The fraction of time that scholars spend responding to administrative tasks has risen dramatically over the past 20 years, leaving ever-decreasing time for the true work of the scholar—the generation and transmission of new knowledge.

How can these challenges be addressed? How can the prospects for scholarship be improved? How might the Internet be used to transform scholarship? Can we create a world in which finding collaborators becomes effortless; storing, retrieving, analyzing, and visualizing information about one's work and the work of others is a natural by-product of the work; and administrators can access information about the work at any time, without the need to interrupt the work of the scholar? We seek to create this new world of scholarship.

In this book, we will introduce VIVO, a new approach to representing and using information about scholarship. VIVO represents a general shift in the use of the Internet—from a collection of "websites" providing "pages" to a Semantic Web of data providing timely information about scholarship around the world. Through this emerging information infrastructure, new applications will emerge to address the issues faced by the scholars and science administrators of today.

1.2   RESEARCH DISCOVERY AND EXPERT IDENTIFICATION

Let us begin by envisioning a world in which scholarship is supported by data and tools. What would such a world contain?

First, there would be information. The information would be timely and accurate. It would be well defined—ambiguous elements would be clarified and redundant elements either removed or clearly discernible as redundant. The information would be linked—that is, if Harvard University is awarded a grant, the grant would be linked to Harvard and Harvard linked to the grant. Linking supports various tools and processes, but it also represents a degree of clarity in the representation of the information. We are confident that "this" grant links to "that" university. The information would be represented in known formats and transmitted using known technologies. This enables participation and world-scale usage. And finally, the information would be openly and publicly available. Scholarship includes both the development of new knowledge as well as its transmission. Our desired state is one in which both activities are visible and consumable by all.

An open, accurate, well-defined, well-represented information infrastructure available to all provides fertile ground for tool developers. The economics of tool development will change from one in which proprietary data representation, access control, and storage are eliminated in favor of common representation and sharing, greatly lowering tool development costs and easing adoption.

A primary use case for this new information is research discovery, broadly defined as processes for learning what is known and what has been discovered over time in one's own, or in other, disciplines. Research discovery is made possible by the new information infrastructure. Knowledge representation, visualization, and linking improve our ability to understand our disciplines, keep up with new developments, and avoid redundancy and rediscovery leading to improved scholarly efficiency and efficacy.

A second use case broadly enabled by the information infrastructure is expert finding [2]— identifying and engaging experts whose scholarly work is of value to one's own. To find experts, one needs rich data regarding one's own work and the work of potential related experts. Relevancy algorithms can then operate on the data to identify expert candidates for collaboration. Communication channels such as email and Skype can be used both indirectly—verifying potential collaborators through social networks—and directly—for engaging potential expert collaborators. Expert finding is of increasing importance in many disciplines. Multi-disciplinary and inter-disciplinary investigation is increasingly required to address complex problems. Expert finding is also useful for identifying competitors as well as job candidates, reviewers, or panelists. Students and trainees can use expert-finding tools to identify potential advisors and mentors.

The new information infrastructure is being built. Linked Open Data [3] is a world effort to make data available in a simple, common format. Resources, such as data.gov [4], provide linked open data for reuse for a new approach to the support of scholarship.

1.3   THE SEMANTIC WEB

The Internet began as a means of sending messages between computers [5]. Early protocols and applications, such as FTP and telnet, were augmented with email and NFS. The breakthrough came in 1992 with the introduction of HTTP and HTML by Tim Berners-Lee [6], resulting in the World Wide Web. Web technologies bloomed rapidly, with CGI, Java, and JavaScript quickly following.

Data representation took a revolutionary step forward with the introduction of XML [7], followed by a second breakthrough, again by Tim Berners-Lee, with the introduction of the Semantic Web in 1998 [8]. Semantic Web technologies, including RDF and OWL, provide the fundamental opportunities developed by VIVO.

The Semantic Web continues to use the Internet for physical transport and HTTP for messaging. But rather than pages of HTML code, the Semantic Web provides data in the form of RDF, a fundamentally new information-representation paradigm [9]. All information in RDF is represented as "triples" consisting of a subject, a predicate, and an object. The object of every triple can be a value or a reference to another element. For example:

x isA Faculty-member

x hasName "Michael Conlon"

x facultyAppointmentIn y

Faculty-member isA Employee

From these triples we see there is a faculty member named Michael Conlon in a department *y*. We can infer (the Semantic Web supports inference) that Conlon is an employee. Figure 1.1 shows a simple use of triples for representing scholarship.

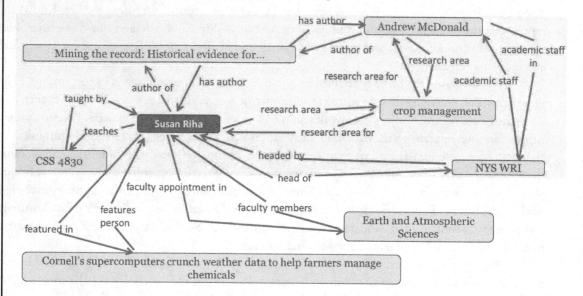

Figure 1.1: Sample relationships between a faculty member, Susan Riha, and elements of the world of scholarship.

The Semantic Web represents information using an extensible set of predicates defining relationships between entities. These can be modeled using the web ontology language OWL [10]. Figure 1.2 shows a simplified view of the types of entities and predicates representing domains of interest to scholars. VIVO has focused on scholars and works. Others have focused on resources [11], while still others have focused on funding and data [12].

Using Semantic Web approaches to modeling and sharing data, we can represent scholarly activity to support the work of scholars: see the sample schema of entities and their relationships in Figure 1.2.

1.4 VIVO: A SEMANTIC WEB INFORMATION INFRASTRUCTURE FOR SCHOLARSHIP

VIVO is an open-source, Semantic Web application used to manage an ontology and populated with linked open data representing scholarly activity. VIVO provides its users with faceted semantic search for expert and opportunity finding, rich semantic linking for research discovery, and profiles of people, organizations, grants, publications, courses, and much more. In April 2012, the Cornell University

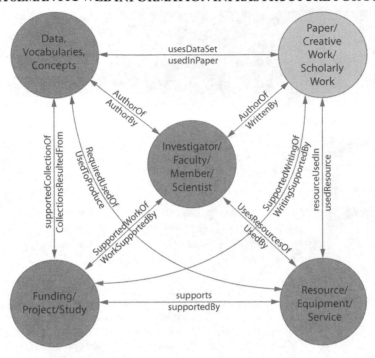

Figure 1.2: Some domains of representation for scholarship.

VIVO contained profiles for 16,231 people, 19,847 organizations, 15,660 courses, 28,839 academic articles, and 6,162 grants. A sample faculty profile is shown in Figure 1.3.

VIVO information is public. One does not need to log in or be otherwise authorized to view the data in VIVO. Think of VIVO as a radio tower—it provides information to anyone who accesses the URL of (turns the radio dial to) the VIVO server. Most university implementations use a simple naming convention of vivo.*name*.edu, where *name* is the name of the university on the web. So, for example, Cornell's VIVO can be found at vivo.cornell.edu, Florida's at vivo.ufl.edu, Indiana's at vivo.iu.edu, and so forth.

Using VIVO is quite simple. As with any other website, one can click on links to be taken to additional information, or use the search box to type search terms and receive links to items in VIVO matching the search terms.

Many VIVO sites permit "self-editing" of data—that is, faculty can edit the material on their profiles. Users log in using a university username and password and are placed on their profile. Simple on-screen controls permit adding, removing, or changing material in VIVO. Most universities find that some of the material in a faculty profile is a matter of public record—federal grants received, for example. Allowing modification of this information might be contrary to university practice and thus disallowed by the university's VIVO.

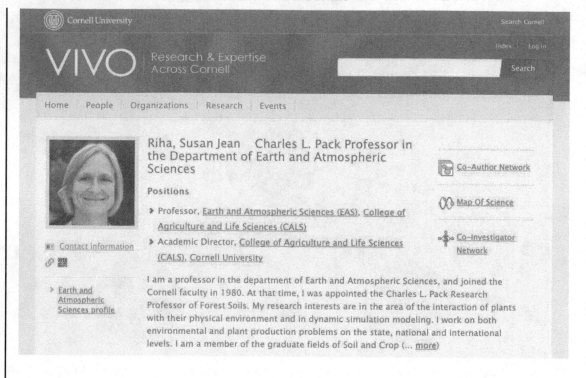

Figure 1.3: The VIVO faculty profile for Dr. Susan J. Riha at Cornell University. Additional information regarding Dr. Riha's work is presented when one scrolls down.

VIVO can present its data as HTML for reading in a web browser. In this way, VIVO is much like other web-based applications. VIVO stores data and presents it on demand based on user clicks on web interfaces.

VIVO can also present its data as RDF (in a variety of RDF representations—N3, Turtle, or RDF-SCHEMA [13], an XML-based representation) for consumption by software. Many VIVO implementations choose to deploy an optional SPARQL [14] end point to allow remote sites to submit queries and receive data in return.

Figure 1.4 illustrates the ways in which VIVO data is provided. A single profile is provided as both HTML and RDF. Software can be written to consume the RDF and transform it into visualizations, analyses, reports, or other desired products. (Writing such software is described further in Chapter 6.) The VIVO community develops software of this kind and discusses it in online forums and presentations at scientific meetings—most notably, the annual VIVO conference.

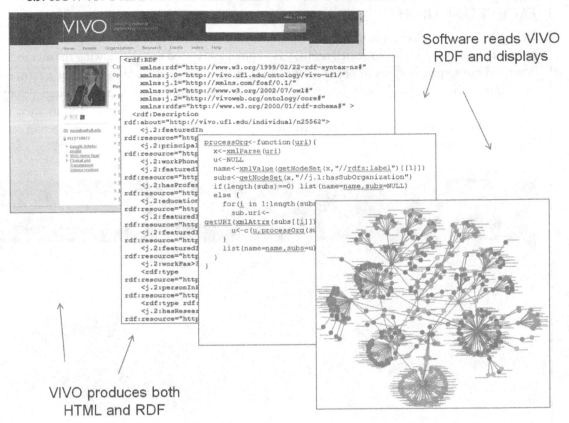

Figure 1.4: VIVO produces both human and machine-readable formats. Software can read the RDF and produce visualizations.

As you may have noticed in the previous descriptions, VIVO implementations vary according to local information provider policy and practice. Implementing VIVO, described further in Chapter 3 of this book, includes information on configuring VIVO to meet organizational standards and needs.

1.5 HOW VIVO ADDRESSES THE NEEDS OF THE SCHOLAR TODAY AND TOMORROW

VIVO provides faceted search, semantic linking, visualizations, and data reuse. Future work can incorporate VIVO data into both existing tools and new tools for scholarship and discovery.

FACETED SEARCH

Unlike Internet search engines that use heuristics and relevance scores to guess what text pages might be of interest to you given your search terms, VIVO provides faceted search, allowing the user to "drill down" to more specific types of objects. One can start with an open-ended search for a term such as "computational biology" (see Figure 1.5). The results include various types of things—people, organizations, papers, grants, courses, and events. If a user then selects "grants," for instance, only grants will appear in the resulting list. This faceting of results is not possible using Internet search engines.

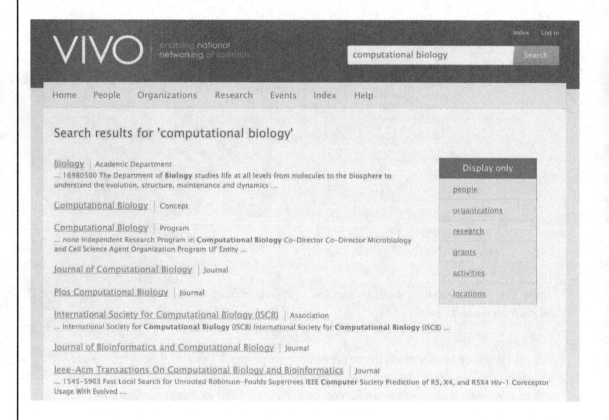

Figure 1.5: A search for "computational biology" returns a wide variety of results—subject areas, programs, papers, grants, and people. The menu on the right provides direct access to subsets of the search results based on the kind of entity.

SEMANTIC LINKING

All links in VIVO have meaning. A grant appears on a faculty member's profile because the faculty member has a defined relationship to the grant. A common relationship is principal investigator. If faculty member x is PI of grant y, then grant y has principal investigator faculty member x (see Figure 1.2). This simple bi-directional semantic linking has powerful consequences for discovery.

For example, suppose you are looking at a faculty member's profile. You see that he or she attended a particular university. You click on the link to that university. On the university's profile page in VIVO you see all grant awards from your university to that university as well as awards that university has made to yours. Following those grant links you find investigators who are funded by that university. Questions such as "who is funded on a sub-contract from university z?" are easily answered by following links.

VISUALIZATIONS

VIVO provides compelling visualizations of person-centric networks, concept networks, and time-based and geography-based displays for understanding the connections between people, organizations, works, and funding over time and place. Chapter 7 describes VIVO visualizations and analysis of VIVO data in much greater detail.

DATA REUSE

VIVO's open display of data in RDF format is ideal for consumption and reuse by new tools. Figure 1.7 shows VIVO data being displayed in the form of "business cards" on a website. The webmaster does not have to maintain data—it is pulled from VIVO in real time as the page is accessed. Software developers can quickly make additional items that reuse VIVO data. Chapter 6 describes extending VIVO and participating in the VIVO development community.

ADDRESSING THE NEEDS OF SCHOLARS IN THE FUTURE

VIVO provides an open information infrastructure for the development of new tools and for the replacement of cumbersome tools of today. These possibilities include the following.

1. Information routing. Given your profile, smart applications could route information of interest to you, rather than have you search for it or rely on email lists that provide too much information of little value.

2. Semantically enabled applications. Calendars, email, and collaboration sites can be semantically enabled to use the information coming from VIVO. When reading an email from an individual, you should be able to know what is publicly known about that individual.

3. Vitae and biosketches. Given the profile data in VIVO, a one-button click can produce a curriculum vitae or biosketch.

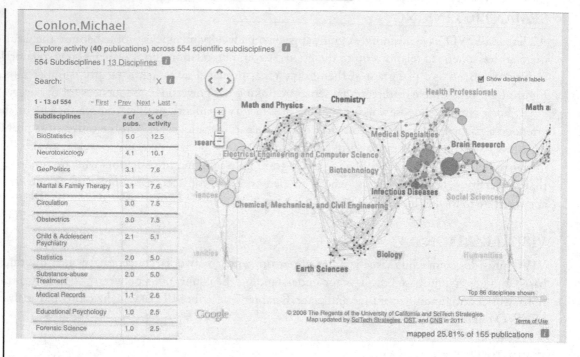

Figure 1.6: VIVO Map of Science for Michael Conlon. From the Conlon profile available at http://vivo.ufl.edu/mconlon, one can select "Map of Science" to generate a concept map displaying recurring sub-disciplines in Conlon's papers.

4. Groups. We all have many groups of people that we interact with—classes of students, research groups and teams, administrative groups and office mates, colleagues and competitors. A semantically enabled mobile application for maintaining your groups would allow you to know what is known, as well as communicate with group members and track changes in public information about them.

5. Bibliography management. VIVO could accept bibliography information from common tools, providing a simple mechanism for import/export and maintenance of high-quality faculty information.

6. Faculty reporting. If all the information about a faculty member is in VIVO, there would be no need for annual faculty summary reports. The university could query the information as needed and make whatever reports they need whenever they need them without resorting to asking the faculty to provide information that is already known. The interconnectivity of VIVO data can also provide new insights into patterns of collaboration among individuals and programs inside and outside the home institution.

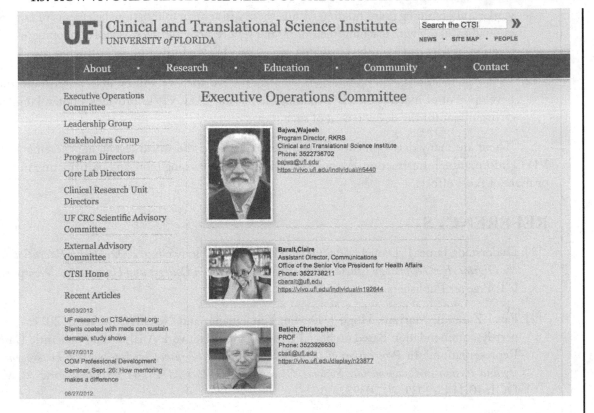

Figure 1.7: VIVO data is used here to produce "business cards" for people on a website. The webmaster merely references the VIVO URL of the desired person to produce the business card.

7. Grant submission and management. Funding agencies could consume VIVO data rather than request faculty to fill out cumbersome and repetitive forms asking for information that is already discoverable. Reuse of data provides added incentives for accuracy and completeness.

8. Course management. Course Management systems could use known data to populate profile information rather than requiring re-entry.

9. Expert finding. Using rich semantic data, our ability to locate relevant expertise for scholarly collaboration will be greatly enhanced.

10. Opportunity finding. Similarly, using rich semantic data, our ability to alert faculty of opportunities relevant to their research and career development will be greatly improved.

11. Next-generation attribution. A growing community of scholars has recognized that contribution to the advancement of knowledge consists of more than authoring well-recognized

works and obtaining grant funding. While these will remain foundational elements of faculty evaluation, along with teaching/mentoring and service, activities such as tool development and creation of data can easily be attributed using the VIVO information infrastructure.

12. "Nano publications." Some futuristic scholars believe that we will be able to attribute the development of individual facts or findings to authors [15]. VIVO provides a potential pilot environment for the development of such ideas.

There are many development efforts underway around the world leveraging the emerging VIVO information infrastructure. We hope you will consider learning more and participating in one or more of these efforts.

REFERENCES

[1] Duderstadt, James J., Daniel E. Atkins, and Douglas van Houweling. 2002. *Higher Education in the Digital Age: Technology Issues and Strategies for American Colleges and Universities.* Westport, CT: Praeger Publishers. 1

[2] Fazel-Zarandi, Maryam, Hugh J. Devlin, Yun Huang, and Noshir Contractor. 2011. "Expert Recommendation Based on Social Drivers, Social Network Analysis and Semantic Data Representation." In *Proceedings of the 2nd International Workshop on Information Heterogeneity and Fusion in Recommender Systems, Chicago, IL, October 23–27*, 41-48. New York: ACM. DOI: 10.1145/2039320.2039326 3

[3] Linked Data—Connect Distributed Data Across the Web. http://linkeddata.org 3

[4] DATA.gov: Empowering People. http://data.gov 3

[5] Krol, Ed. 1994. *The Whole Internet User's Guide and Catalog.* 2nd ed. Sebastopol, CA: O'Reilly Media. 3

[6] Berners-Lee, Tim. 1989. "Information Management: A Proposal." The World Wide Web Consortium (W3C). http://www.w3.org/History/1989/proposal.html 3

[7] "Extensible Markup Language (XML)." 2012. The World Wide Web Consortium (W3C). Last modified January 24. http://www.w3.org/XML 3

[8] Berners-Lee, Tim. 1998. "Semantic Web Road Map." The World Wide Web Consortium (W3C). Last modified October 14. http://www.w3.org/DesignIssues/Semantic.html 3

[9] Allemang, Dean, and James Hendler. 2011. *The Semantic Web for the Working Ontologist.* 2nd ed. Waltham, MA: Morgan Kaufmann. 3

[10] McGuinness, Deborah L., and Frank van Harmelen, eds. 2004. "OWL Web Ontology Language Overview." The World Wide Web Consortium (W3C). Last modified February 10. http://www.w3.org/TR/owl-features 4

[11] Eagle-i. http://www.eagle-i.net 4

[12] DataCite Home Page. http://datacite.org 4

[13] Brickley, Dan, and R.V. Guha, eds. 2004. "RDF Vocabulary Description Language 1.0: RDF Schema." The World Wide Web Consortium (W3C). Last modified February 10. http://www.w3.org/TR/rdf-schema 6

[14] Prud'hommeaux, Eric, and Andy Seaborne, eds. 2008. "SPARQL Query Language for RDF." The World Wide Web Consortium (W3C). Last modified January 15. http://www.w3.org/TR/rdf-sparql-query 6

[15] Mons, Barend, Herman van Haagen, Christine Chichester, Peter-Bram 't Hoen, Johan T. den Dunnen, Gertjan van Ommen, Erik van Mulligen, Bharat Singh, Rob Hooft, Marco Roos, Joel Hammond, Bruce Kiesel, Belinda Giardine, Jan Velterop, Paul Groth, and Erik Schultes. 2011. "The Value of Data." *Nature Genetics* 43 (March): 281–283. DOI: 10.1038/ng0411-281 12

CHAPTER 2

The VIVO Ontology

Jon Corson-Rikert, Stella Mitchell, and Brian Lowe, *Cornell University*
Nicholas Rejack, *University of Florida*
Ying Ding and Chun Guo, *Indiana University*

Abstract

VIVO does much more than add semantic tags or controlled vocabulary terms to content in a traditional researcher profiling system. By storing all data internally using types and relationships defined through Semantic Web standards, VIVO enables sophisticated data browsing, query, and visualization capabilities in a single package within one institution or across a network. Through Linked Data[1] standards, information in one VIVO can also be connected directly to any other store of semantically encoded information anywhere on the Internet.

This chapter will focus on semantic representation through ontologies and how this form of representation undergirds the entire VIVO project, including the main VIVO application, the VIVO Harvester, and the indexing and search tools built to demonstrate the potential of a distributed network of VIVO applications interoperating with other tools that take advantage of the VIVO ontology for data interchange. We will discuss the VIVO ontology in the overall linked data context, including new requirements not typically addressed by standalone applications. We will provide an overview of what VIVO does and doesn't try to model, discuss our processes for sustaining and extending the ontology over time, and review many of the choices we have made in establishing the ontology, including how the ontology affects the VIVO application and how the requirements of making a working application have influenced our approach in ways we have not always anticipated.

The chapter then closes with an example of collaboration beyond the VIVO community and a brief look at challenges and future directions.

Keywords

ontology, RDF, OWL, linked data, standards

[1] http://linkeddata.org

2.1   INTRODUCTION

VIVO is an application with many ambitious goals: to provide an institutional platform for searching and browsing data about researchers and their activities; to represent a coherent view of an institution's research in great detail to the outside world; and to supply that data in turn to other applications.

One of the motivating goals of the Semantic Web as a whole is to allow new applications to emerge more rapidly by leveraging data that may already have been gathered for different purposes, very likely from multiple sources. Semantic Web standards promote interoperability of data based on the precept that no application is completely self-contained; an investment in data for one purpose can realize unforeseen benefits when that data can be reused for other purposes through other applications. Having data already available reduces the risk and the cost of developing new applications.

While Semantic Web data formats are simple, their successful application in practice requires the adoption of one or more ontologies defining what is represented in the data and how the different entities represented relate to one another. This chapter discusses the basics of ontologies and the VIVO ontology's function within the VIVO application as well as its critical role in permitting data about researchers to be aggregated, analyzed, or visualized independently of VIVO itself. In fact, an increasing number of other applications have been adapted to output data conformant to the VIVO ontology.

The VIVO ontology is very much a living, breathing thing—as VIVO grows in its reach, it is essential to have established processes for addressing new adopters' needs in ways that support consistent communication of meaning at the level of a regional consortium, a country, or worldwide.

2.2   SEMANTIC TECHNOLOGIES

2.2.1   RATIONALE FOR USING SEMANTIC STANDARDS

We have chosen the Resource Description Framework (RDF) standard for data encoding[2] and the Web Ontology Language[3] (OWL) as the logical structure for the VIVO application because of their wide adoption and because of the increasingly rich and diverse set of associated tools available. These tools span the gamut from simple browser and content management plugins to data manipulation tools, triple stores, the SPARQL query language,[4] and a selection of reasoners. We can also take advantage of previous ontology work as well as ongoing projects to create and refine ontologies in a wide range of academic and scientific domains.

2.2.2   RDF AND OWL

RDF is a simple, flexible data model for representing information on the World Wide Web, and it is intended to facilitate merging of disparate data sets. A VIVO dataset is composed of (RDF)

---

[2]http://www.w3.org/rdf
[3]http://www.w3.org/TR/owl-features/
[4]http://www.w3.org/TR/sparql11-query/

triples, and those triples are given meaning by the VIVO ontology. The web ontology language (OWL) is designed to allow meaning to be encoded in a form that computers can understand and manipulate, most elementally by defining types of things (classes) and the relationships between them (properties).

An RDF triple is composed of a subject (a thing you are saying something about), a property (the nature of what you are saying about the subject—e.g., what its color is, or who its sister is), and an object (the content you are conveying—e.g., red, or Sally). In most VIVO triples, the subject is the member of a class defined in the ontology, or an *individual* of that class, and is usually identified by a URI, or a unique identifier encoded as an address on the web.

A property represents a relationship between one subject and something else in the object position of the triple. The object may either be another individual, capable itself of being the subject of other triples (e.g., Sally), or a simple literal value (e.g., the string "red"). Properties between individuals connect the nodes in the RDF graph of data; an individual can be the subject and/or the object of any number of RDF triples. Triples are also commonly referred to as *statements* or *assertions*.

The classes in an ontology have relationships to each other via the OWL *subclass* and *equivalent class* relationships. An ontological hierarchy represents "is-a" relationships, which means that each member of a child class is also a member of the parent class. The class hierarchy, or taxonomy, moves conceptually from general to specific, and everything about a parent is also true for its children.

2.2.3 LINKED DATA

The terms Linked Data and Linked Open Data refer to the open interchange of data on the Internet using standard communication protocols and identifiers—the familiar HTTP protocol used for web page requests and responses, the RDF data model, and uniform resource identifiers or URIs to uniquely specify each element of data. For linked data to conform to accepted conventions, these URIs must be directly accessible on the web as URLs or uniform resource locators.[5] Not all linked data is open or publicly available on the web without restriction.

2.2.4 FEATURES OF SEMANTIC MODELING

This section describes five notable features of modeling for the Semantic Web using OWL that sometimes require a different paradigm of thinking than when working with more traditional modeling environments.

Reasoning: The OWL Web Ontology Language has a formal semantics for expressing *axioms*, or statements that are assumed to be true. An OWL axiom can often be expressed with a single RDF triple, while some types of axioms—especially those defining OWL classes—require small sets of triples. In turn, these axioms provide the basis for making further inferences about an ontology and the data populating that ontology. A software program (i.e., a reasoner) can add additional data statements (triples) to the knowledge base, based on triples that already exist.

[5]http://linkeddata.org

No Unique Name Assumption: In logic languages, the Unique Name Assumption means that different names always refer to different individuals in the world. OWL does not make this assumption, but it does provide a vocabulary with which to indicate two individuals are the same or different from each other.

Constraints: OWL is designed to support inference; it is not a schema language for applying integrity constraints.   For example, if a Position is defined as being in a single organization, linking it to the URIs of two organizations will not directly cause a constraint violation: instead, the reasoner will conclude that the two organizations must be the same. Similarly, if two organizations are asserted to be the same, both would be inferred to be connected to the position. Reasoning works in both directions, and thus allows us to fill in gaps based on partial information regardless of which part of the information we may start out with.

Only if the URIs of the two organization URIs were already asserted to be different from one another (or if there were other axioms causing them to be inferred to be different) would the reasoner report that the ontology is *inconsistent*. Inconsistency is, for practical purposes, an error state that prevents any useful reasoning from being performed.

Open World Assumption: The absence of a statement has a very different meaning in Semantic Web languages such as OWL than in constraint-based languages or database schemas. Under the open world assumption, the absence of a statement does not mean the statement is false. For example, if the Faculty Member class is defined in such a way that its members must have at least one Teacher Role, the absence of an asserted statement about a teaching role for any given faculty member does not cause an error. OWL allows a reasoner to infer that a teacher role exists even if the available data does not state its URI. The URI may be added through merger with another Semantic Web dataset or it may be left forever unspecified.

Monotonic Inference: OWL reasoning is monotonic, which means that if a statement is inferred from a set of axioms, adding an additional axiom cannot remove that inferred statement. To invalidate the inference would require at least one of the original axioms to be removed.

2.3   DESIGN OF THE ONTOLOGY

In 2003, the VIVO application was conceived at Cornell for broad application in the life sciences and by 2007 it had expanded to encompass the whole university in scope. Therefore, it has never sought to represent any single academic domain in great detail. The ontology focuses in large part on commonalities rather than on differences in academic domains—the people, activities, organizations, affiliations, events, processes, and services that interact through well-understood relationships across most academic disciplines. This is not to say that there are not disciplinary differences in what needs to be represented—from visiting critics in architecture to performances in theater to clinical studies in medicine.

2.3.1 GOALS

The VIVO ontology is a unified, formal, and explicit specification of information about researchers, organizations, and the scholarly activities, outputs, and relationships that link them together. The ontology forms the core of the VIVO application that in turn provides the means by which data conforming to the ontology can be created and disseminated. The ontology must have its own integrity independent of the application because the data must have coherence and value outside the application.

Therefore, an ontology is more than a data model or schema for one application—it must be seen more broadly as a framework for data, rooted in standards, that conveys meaning in a coordinated, linked network of statements defining and relating many different freestanding pieces of information. Because the data must be coherent both inside and outside the application, not just within, the ontology must contain (or directly reference) its own definitions.

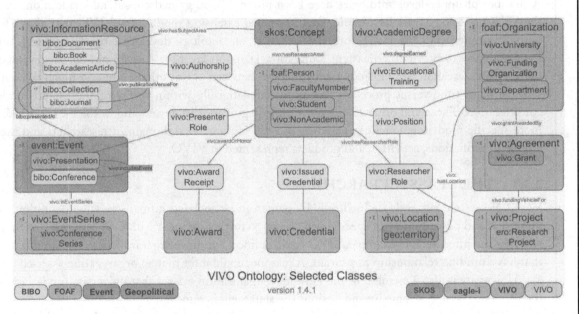

Figure 2.1: Selected major classes in the VIVO ontology showing their source ontologies and the "context nodes" (in light blue) carrying temporal and other information specific to an individual relationship. A full ontology listing appears in the Appendix.

2.3.2 INDEPENDENCE

The VIVO project is not about abstract encoding of knowledge. The intention with each new adopting institution or group is to populate VIVO with real-world data, including all its messiness. Translating the continuously evolving three-dimensional world into useful surrogates for information discovery and exchange will of course sacrifice much of the richness of the original. But we as humans have implicitly understood sophisticated concepts, such as substitutions of maps for spatial reality, even in cultures without written communication.[6]

Selecting how to collapse the dynamic fullness of the world into a lightweight and intuitive connective framework requires balancing many independent factors, with inevitable compromise. The most frequent modeling choice involves balancing detail and/or specialization against the utility of common experience. The VIVO ontology is overtly trying *not* to model any one institution or discipline at a level that precludes integration into a larger body of knowledge. For VIVO, the generality versus specificity balance falls at a level just above domain ontologies that are typically developed to represent knowledge in a single discipline or field of specialization.

Moreover, VIVO is also *not* attempting to model everything in the world or to provide an umbrella ontology connecting representations of multiple domains in a single conceptual theme. A number of upper-level ontologies have been proposed and gained considerable adoption,[7] and the VIVO ontology team has developed an optional extension module for VIVO that aligns the ontology underneath the Basic Formal Ontology,[8] an ontology developed by Barry Smith and Pierre Grenon to provide a consistent foundational support layer for domain ontologies developed for scientific research. BFO introduces distinctions between entities that endure in much the same state through time versus processes that unfold dynamically within the temporal bounds under consideration. These and other distinctions provide guidance in developing ontologies addressing more specific modeling domains, including the still fairly general domain of researchers and their attendant affiliations, activities, and products represented in VIVO.

2.3.3   THE CLASS HIERARCHY

VIVO classes are currently defined only by names and associated human-readable definitions. A more rich and powerful way of specifying a class is to further define it with OWL statements that a reasoner can interpret. For example, VIVO could define a Funding Organization as any organization that has a funding relationship to a Grant. A reasoner could infer that an organization asserted only as a University is also a member of the Funding Organization class if that university funds a grant; the inference would become invalid again if the statement is removed.

The VIVO ontology defines a relatively small number of types for people. In part, this reflects a bias for enabling sharing of data among institutions, where subtle distinctions often important at the local level become blurred by differences in terminology. While precise academic rank may be important to reflect in a person's title, an assistant professor does not typically conduct different activities or enter into different relationships at the institution than an associate professor, and the VIVO ontology does not need to distinguish assistant professor and associate professor as person types (classes) distinct from faculty member.

---

[6]Jaime Morrison, "Polynesian Stick Charts," *The Nonist* (blog), August 23, 2003,
 http://thenonist.com/index.php/thenonist/permalink/stick_charts/.
[7]http://en.wikipedia.org/wiki/Upper_ontology_(information_science)
[8]http://www.ifomis.org/bfo

One could theoretically represent all *class* distinctions via the presence or absence of *properties*—for example, a Teacher is any Person currently having a "teacher of" relationship to a course, and a Student anyone with a "student in" relationship to that course, regardless of the classification on their ID card. In practice, the VIVO ontology distinguishes 13 types of persons, most reflecting ongoing distinctions that remain relatively constant and are associated with notably different relationships in the academic and research community, such as the distinction between faculty member, non-faculty academic, non-academic staff, student, and emeritus.

The full VIVO ontology class hierarchy and the object and datatype properties associated at each level are included for reference in the Appendix.

2.3.4   MODELING PRINCIPLES

Ontology Reuse

Reusing ontologies not only saves effort but also increases the inherent compatibility of data that may be developed in different applications, at different times, and in different places.

Following the MIREOT principle, or Minimum Information to Reference an External Ontology Term,[9] VIVO selectively reuses classes and properties from external ontologies, especially from ontologies that are well documented and in wide usage on the Semantic Web. Rather than minting a VIVO URI for the same class or property and publishing a mapping file that must be maintained, or importing an entire ontology when only part of it may be relevant, VIVO uses the URI and definition of the class or property as published in the original ontology. This practice reduces the burden of mapping VIVO data from one ontology to another on export and reinforces the self-documenting nature of linked data—if a new class or property is encountered, it is expected that a definition will be available via an HTTP request to the URI of the class or property.

We have three general scenarios when importing classes and properties from other ontologies. At times, VIVO adds additional detail, as in providing subclasses of foaf:Person and foaf:Organization. In other cases, the VIVO ontology imports a substantial portion of the Bibontology while adding a relatively small number of additional classes and properties using the VIVO namespace.

As the VIVO ontology is extended to address requests for new types of content or new relationships, the expectation will be that an ontology may already exist that meets all or most of the additional requirements. In this third case, VIVO will likely expand only as far as needed to bridge into the domain represented by the new ontology, usually by selecting a relatively small subset of classes and/or properties from the ontology in question and allowing adopting institutions to import further classes or properties as needed in the local context.

---

[9]http://obi-ontology.org/page/MIREOT

Design for Linked Data

The VIVO ontology describes data that are intended not only to provide content for the VIVO software application, but also to be shared openly on the web as linked data and applied to any number of purposes.

To maximize data discovery and promote interoperability, the VIVO ontology strives to reuse established linked data ontologies wherever possible, including Friend of a Friend (FOAF)[10] and the Bibliographic Ontology (BIBO).[11]

VIVO must, however, balance the desire for maximum reuse of existing Semantic Web vocabularies against the goal of being able to model researchers and their activities at a level of detail that supports meaningful communication across diverse institutions and an aggregation of rich data over time. The Dublin Core Metadata Initiative (DCMI)[12] defines a number of widely used terms, such as "creator" and "contributor," that are frequently found in RDF linked data. The VIVO ontology, however, includes a richer notion of authorship that includes author order and provides a place to attach additional detail about a particular individual's relationship to a publication. Consequently, VIVO defines a context node to represent authorship instead of including Dublin Core's creator property.

In other cases, the VIVO ontology provides "shortcuts" to mask details of other ontologies. For example, the Bibliographic Ontology links individual journal articles to nodes representing the journal issues in which they are published. In turn, these are linked to journal volumes, which then lead to nodes representing the journals themselves. While this level of detail is important for modeling the complete publication process itself, it is not expected to be so useful in most VIVO installations. Therefore, VIVO introduces the property "hasPublicationVenue" to link an article to a URI in VIVO for the journal in which it was published, providing a short path for applications crawling linked open data to discover which authors have published in which journals.

Context Nodes

Relationships can be direct and simple or more nuanced. Many relationships have attributes in their own right that are important to represent and associate with the correct context. A professor is a member of a department, but that membership has temporal bounds—an associated university and working title, and perhaps additional attributes such as an honorary endowed chair title. A position is usually associated with employment in an organization continuing over significant time, while roles more typically represent a person's involvement in shorter activities occurring within a narrower time span, such as the duration of an academic semester or a conference. The ontology defines positions and roles quite generally, while providing options for more specific child classes when warranted by meaningful distinctions in data; nothing in the ontology prevents a role from lasting multiple years or a position having a much shorter duration.

---

[10]http://www.foaf-project.org/
[11]http://bibliontology.com/
[12]http://dublincore.org/

Much of the richness of VIVO is expressed through this consistent pattern of relationships that themselves have attributes and function as "context nodes" connecting people and organizations with other first-order entities including grants, publications, projects, events, or research resources such as labs, equipment, techniques, studies, and protocols. In addition to positions and roles, context nodes represent educational training and the receipt of awards and distinctions, and credentials; each context node reflects a single instance of a relationship between two independent things each likely to have many independent relationships.

Extensible Design

In trying to remain as useful as possible across domains, and to allow the VIVO application the flexibility to evolve, the VIVO ontology attempts to model relevant types and relationships "as representations of the reality that is described by science."[13] The data in VIVO are intended to be independent of any single application domain or function and should not be constrained or distorted to fit immediate operational or transactional needs. In part, this choice acknowledges the cumulative value of institutional investments in data in contrast to more ephemeral investments in software, but it also reflects a conscious desire to provide meaningful and lasting structure in data to help assure portability. Convenience in mapping from the data source and simplicity for the application cannot be ignored, but compromises may limit the value of the data.

As a case in point, the VIVO ontology community has had many discussions about how to store authors of publications when nothing is known about them except the name included in the bibliographic metadata, often with only a first initial beyond the last name and no email address or other unique identifying information. It would be convenient to store all authors in a single delimited text string, with individual authorship relations to any known authors or authors at the home institution. Another approach would be to create authorship relations but simply store the author name string on each unknown relationship. The VIVO ontology instead models every author as a foaf:Person based on the assumption that more information may become available in the future to supplement what little is already known. One team at the 2011 VIVO hackathon event[14] found numerous common author references in searching across several VIVO sites, and by having foaf:Person records, could supplement existing information with full contact information from another VIVO.

2.4   RELATIONSHIP TO THE APPLICATION

2.4.1   ONTOLOGY AS DATA MODEL

VIVO's flexible and extensible data model allows it to present an integrated view of people and their activities across an organization, featuring links among them and connections to other people as well as their professional information by using a network graph structure to most naturally represent a

---

[13] Barry Smith and Werner Ceusters, "Ontological Realism: A Methodology for Coordinated Evolution of Scientific Ontologies," *Applied Ontology* 5, 3-4 (2010): 139-188, doi:10.3233/AO-2010-0079.
[14] http://sourceforge.net/apps/mediawiki/vivo/index.php?title=2011_VIVO_Hackathon

real-world network of relationships. There are many ways a person's expertise may be discoverable, including talks, courses, and news releases as well as through research statements or publications listed on their profile, resulting in the creation of implicit groups of people based on a number of pre-identified, shared characteristics.

The focus on a high-quality, standardized ontology with the application built on top as a population and delivery mechanism for the information protects long-term investment in the data. It also enables independent producers, such as Harvard Profiles,[15] Loki,[16] and Digital Vita,[17] to offer RDF data formatted according to the VIVO ontology and contribute to a national (or international) network of researcher information. Sharing a common representation format allows data from different institutions to be searched and aggregated in a meaningful way.

2.4.2   REASONING

One of the benefits of an application that can reason is that users entering new data can make minimal assertions based on what they know at the time that new information is added to the application; later independent assertions may cause the reasoner to infer additional class memberships. As demonstrated in our example from Section 2.3.3, a university may at times provide subcontracts on sponsored research to other institutions. When these grants are added, the university may be inferred to be a Funding Organization in addition to its existing asserted classification as a University. The ability of a reasoner to infer additional statements about existing data has the potential to provide much additional value to a Semantic Web application.

The VIVO application currently depends mainly on class hierarchy reasoning based on subclass and equivalent class relations. When an individual is asserted to be a member of a given class, the reasoner adds additional statements to show that it is also a member of any parent or equivalent classes. This allows data to be entered at a lower level of the class hierarchy (e.g., InvestigatorRole) and to be queried and discovered at a higher level (e.g., Role). This principle also applies when aggregating VIVO data involving an institution-specific ontology extension defining a new subclass of a VIVO core class such as Grant. Through inference, all such grants will be recognized as members of the core Grant class and found as such in a search. Class hierarchy reasoning is also used to organize content on the index and menu pages, to provide facets to search results, and to ensure that properties defined for parent classes are offered for population during child class editing.

Most object properties in VIVO are defined to be bi-directional through the explicit declaration in the ontology of an inverse property opposite in meaning. Bi-directional relationships allow users in the VIVO application to navigate from a person to a related department or grant while also supporting lists of department members on department pages or investigators on grant pages. Similarly, if the Biology Department is a "subOrganizationOf" its parent College of Life Sciences, the College "hasSubOrganization" Biology Department. An end user can easily see and navigate

---

[15]http://profiles.catalyst.harvard.edu/
[16]https://www.icts.uiowa.edu/confluence/display/ICTSit/Loki-Vivo+Alignment
[17]https://digitalvita.pitt.edu/

to contextual information from wherever they first arrive in VIVO, most commonly from a search engine. If one direction of a property with a declared inverse has been populated, VIVO will add the inverse statement.

VIVO's embedded reasoning service currently consists of a combination of Pellet[18] for reasoning on the class and property hierarchies and a custom-built reasoner that maintains certain types of inferred statements about individuals. The custom reasoner is designed to use very little memory and to update inferences rapidly as data is added to or removed from VIVO. Taking full advantage of reasoning for classification and connection of data in additional ways will require augmenting VIVO's current capabilities either by enhancing the built-in reasoner or by making it possible to take advantage of external reasoners designed to work with large-scale instance data. Both of these approaches are avenues for future development.

2.4.3   COMMON IDENTIFIERS FOR SHARED INDIVIDUALS

While it is easy to promote discovery of linked data connections within a single data source, challenges arise in trying to aggregate data and draw connections across organizations. Making meaningful RDF links across institutional boundaries depends either on the use of common URI identifiers or on processes to discover which sets of different URIs actually refer to the same resource.

If two different VIVO installations want to publish RDF triples asserting that certain academic articles were published in a given journal, they need to use URIs to identify that journal. The quickest and technically simplest approach is for each VIVO installation independently to make up a new identifier by minting a new URI. On the Semantic Web, this is perfectly valid. Without additional information, one cannot assume that two different Semantic Web URIs refer to two different things since there is no unique name assumption in effect. Note that it is typically not appropriate to use the URL of the journal's public website as its identifier in RDF triples: the website URL identifies an electronic web resource *about* the journal, not the journal itself.

If multiple URIs for a journal are published, they may be discovered to represent the same thing based on certain properties they share in common. There may be RDF triples asserting, for example, that each of the URIs is associated with the same ISSN number. We can explicitly state that these URIs represent the same journal through a special sameAs property defined in the OWL language. Adding triples using sameAs allows linked data harvesters to follow the connections between different sets of data about the same resource. It also allows reasoners to conclude that any statements asserted about one URI are also true of all URIs that are sameAs the original.

Where the number of instance values in a class is relatively small and stable, as with Academic Degree and Country, the VIVO ontology has included individuals to be shared across all VIVO deployments and other software compatible with the ontology.

---

[18]http://clarkparsia.com/pellet/

2.4.4   EXTERNAL CONTROLLED VOCABULARY REFERENCES

Controlled vocabularies, taxonomies, and thesauri have been created in order to standardize terminology in various disciplines. These efforts, such as the National Library of Medicine's Medical Subject Headings (MeSH),[19] the UN Food and Agriculture Organization (FAO)'s AGROVOC,[20] or the Library of Congress Subject Headings (LCSH),[21] may represent decades of work by standards bodies or other collaborative initiatives. The interoperability of Semantic Web data may be greatly enhanced by including references to terms from vocabularies that have been adopted by their respective communities, and this task is made easier as more vocabularies are published as linked data with stable URIs for each term.

These terminologies, while typically hierarchical and expressing some notion of cross-reference between terms, may have semantics significantly different from a typical OWL ontology. It is still useful to associate a Semantic Web resource with an appropriate term from a controlled vocabulary. For example, a person can be associated via the "hasResearchArea" property with a topic that is the focus of their research.

The VIVO application supports querying external web services to discover appropriate vocabulary term URIs and linking them to resources in VIVO, without the need to download or import entire vocabulary systems wholesale. This approach gives VIVO adopters the flexibility to select the terminology most appropriate to their field of research while making vocabulary terms readily discoverable by applications that harvest linked data. In the future, for example, search engines may rely on these terminology annotations to enhance precision or to sort results more accurately by relevance.

2.4.5   MIGRATING INSTANCE DATA

It can be challenging for any application to balance requests to encompass new data modeling capabilities against the difficulty of keeping existing data up to date with changes. In addition, the ontologies imported by the VIVO core may themselves change.

In the VIVO life cycle, data population and ontology development have proceeded in parallel. This has allowed for progress on both fronts and has also allowed results from each effort to feed into the other in the context of the VIVO application. The type and structure of data available bring up requirements for what needs to be modeled in the ontology. Ontology features added as a result of user facing requirements for what should be shown on a profile page or available as a search or query inform decisions as to what kind of data acquisition activities should be pursued. As data are populated and viewed via the application, new requirements are raised for what data are needed and how that data should be modeled.

As a result of the ontology design changing in conjunction with existing data stores at multiple deployment sites, we developed a largely automated process to bring a deployment's knowledge-base

---

[19] http://www.nlm.nih.gov/mesh/
[20] http://aims.fao.org/website/AGROVOC-Thesaurus/sub
[21] http://www.loc.gov/aba/cataloging/subject/

instance data into conformance with the current version of the ontology. This is accomplished with software that on system startup detects a knowledge base that has not been examined for updates relevant to the current version, and performs necessary modifications. These modifications may include name (URI) changes for a class or property, removal of a class or property, or structural changes in how individuals are related to each other, including sometimes the addition or removal of an intervening individual (context node). The implementation of this knowledge base migration process uses Protégé PromptDiff[22] to generate a table of differences between two ontology versions, and a combination of java code and SPARQL construct queries to effect the changes.

2.4.6   INTEGRATED ONTOLOGY EDITOR

To enable rapid prototyping of changes in the ontology and the evaluation of the impact on the application of adding new kinds of data, VIVO includes a web-based ontology editor. Since most ontology editing functions can be accomplished within VIVO, this simplifies the number of tools to be learned when adopting VIVO. When logged in with sufficient privilege, a user can easily navigate from an individual or a particular piece of information about an individual to the relevant class or property in the ontology editor. Likewise, from the ontology editor there are functions to find all individuals of a class or all assertions involving a particular property. The ontology classes and properties are available as linked data as well as the populated individuals and relationships. Having sufficient functionality to support public-facing, ontology-based applications within a single web platform enables distributed and collaborative cycles of ontology creation, population, review, and revision. The ontology is persisted in a triple store, and an abstraction layer allows a choice of knowledge base stores.

2.5   EXTENDING THE ONTOLOGY

As the core VIVO ontology represents elements we would expect to find at any institution, it should necessarily be kept simple and relatively stable. Simplicity enables adaptation at many different kinds of institutions and stability ensures that sites running different versions of the ontology will not find mismatches across their data.

Implementers are encouraged to extend the ontology in a way that offers more detail but which still maintains interoperability with the larger VIVO ontology structure. This facilitates data interchange and common search over aggregated data. It is part of the design of VIVO to provide a framework for extensibility under a common core ontology in the context of an application that can automatically accommodate the extended ontology in its data entry, web display, and search capabilities.

The ontology has been developed with academia in mind, so local extensions may be required for institutions far outside this realm. Just as requirements vary across institutions and domains for

---

[22]http://protege.stanford.edu/plugins/prompt/PromptDiff.html

the information modeled, so also do the requirements vary across countries, especially when the data is used for reporting purposes.

2.5.1 MODELING GUIDELINES

When creating a local extension, one of the first decisions is whether the data to be represented will be a primitive string or numeric value (such as a local unique identifier for a person) or whether it will represent an individual with an independent existence, such as a Courtesy Faculty type of a person. The first requires creating a new datatype property to assert about a person, whereas the second requires a new class. In addition, connections between members of classes are enabled by creating new object properties. For example, an institution may wish to specify that Libraries can contain Collections and create a new object property to link the two.

VIVO core reuses properties and classes from several different ontologies, and investigating existing ontologies for reuse when adding extensions is similarly recommended. It is easy to fall into the trap of over-modeling. An institution may have 15 different types of extension offices, but it may be better to simply model them as 1 type if there is no practical need for distinction.

Multiple Inheritance

The VIVO ontology team has taken no *a priori* position for or against multiple inheritance in class definitions—i.e., the assertion that one class is a subclass of more than one other class (excluding classes in a direct line further up the class hierarchy). The VIVO core ontology uses multiple inheritance only minimally, in part because multiple inheritance introduces additional conceptual complexity for humans. Also, since the independently created and maintained parent classes may include axioms that conflict, multiple inheritance can also make the work of computer software reasoners more complex[23] and hence slower, or in the worst case, may result in the failure of reasoning to complete.

Asserting that an individual entity (e.g., a particular organization) is a member of multiple classes is an accepted technique in VIVO and has fewer implications for reasoning, although it may of course result in inconsistencies if any of the relevant classes are defined to be disjoint (have no members in common).

2.5.2 CASE STUDIES

Organizational Hierarchy

One frequent exercise for an institution adopting VIVO is to define an organizational structure to knit together information about a range of people and departments. When doing this, it is important not to confuse an organizational hierarchy with a class hierarchy. Ontological hierarchies represent

---

[23]Johan Dovland, Einar Broch Johnsen, Olaf Owe, and Martin Steffen, "Incremental Reasoning for Multiple Inheritance," in *IFM '09 Proceedings of the 7th International Conference on Integrated Formal Methods, (iFM'09), Düsseldorf, Germany, February 16-19* (Berlin: Springer-Verlag, 2009), http://heim.ifi.uio.no/$\sim$einarj/Papers/dovland09ifm.pdf.

"is-a" relationships, in which the child elements are members of the parent classes. For example, since a department is not a college, the Department class should not be a subclass of College. Instead, individual departments can be linked to colleges using the "subOrganizationWithin" property, which represents a "part of" relationship.

Internal vs. External Individuals

There arose a common need in VIVO installations to showcase individuals directly affiliated with the host institution while exposing related individuals from other institutions only in selected circumstances. For instance, it is useful and important to find that a Cornell professor is a coauthor with researchers from other universities when viewing the professor's publications, but it is not desirable for these external collaborators to be featured in the browser and search interfaces as if they were persons directly affiliated with Cornell.

Although initial instincts and attempts to address this situation were focused on identifying external individuals to *exclude* from those displays, these turned out to be unsatisfactory because of the fact that VIVO is operating in an open world assumption context. Later information may clarify that a person is internal to the institution. By instead identifying individuals that are internal, and by using an institution-specific namespace to do this, we introduce only ontology terms that are semantically meaningful on the larger Semantic Web and when data from multiple institutions are aggregated.

The implementation of this technique is to introduce a class as a local extension at the highest level in the class hierarchy actually needed. For example, Cornell might create a class called cornell:CornellAgent and add assertions that each of its staff members and organizations are members of this class. Since internal is a relative concept, the "internal indicator" class needs to be an extension in a local namespace. These classes will thus continue to make sense when RDF data are combined from multiple institutions—that is, the classes will still contain individuals relevant to their home institutions.

A class representing "internal" individuals rather than "external" works better with the open world assumption, which tells us that we cannot conclude something is not true simply because it is not asserted in an RDF triple. If we consider a Cornell Person to be any person with a Cornell netID, then in the VIVO application we do not want to conclude that an individual without a netID is not a Cornell Person since they may in fact have one of which we simply do not yet have a record. It is therefore more effective to be able to define in the affirmative that a Cornell Person is a person with a netID. If a netID assertion is later added for a person, a reasoner could then automatically conclude that the person is a member of the Cornell Persons defined class.

2.6   VIVO ONTOLOGY COMMUNITY—EFFORT

As an open-source ontology, VIVO has welcomed community contributions both to the core and in the form of optional extensions. However, since the ontologies must align to enable data exchange, the "let every flower bloom" approach has been tempered by consensus-driven top-down decisions made by the ontology group. All of these changes take place in a community-driven, visible process that aims for high responsiveness and sustainability. At the regular ontology phone conference, a new modeling requirement can be raised and discussed by the group to compare and contrast the data representation needs at different sites. A preliminary design may be suggested for either inclusion in the core ontology or as a local extension for one or more institutions. After further research, the change or addition is typically reviewed again in a subsequent phone conference before inclusion into the core ontology or recommendation as a local extension.

With the recent submission of the VIVO 1.4 ontology to the National Center for Biomedical Ontology's Bioportal ontology index,[24] more community feedback is expected.

2.7   LOOKING AHEAD

The VIVO ontology will continue to evolve as the VIVO community needs it to, especially as new types of data are proposed for inclusion in VIVO and need to be modeled. Newly adopting institutions frequently introduce new requirements that may have relevance more broadly, and collaborations with other projects and initiatives bring new ideas and important additional knowledge to the VIVO community.

The U.S. National Institutes of Health (NIH) elected to make a major commitment to the Semantic Web through support of VIVO and eagle-i,[25] a closely related project to model research resources. The eagle-i and VIVO ontology teams have worked to align their respective ontologies at a high level, and this collaboration has been extended through the award of a contract to the Oregon Health and Sciences University for CTSAconnect.[26] This project will create a single integrated semantic framework for information on researchers and research resources, focusing on improved recognition of the many components contributing to research expertise bridging the clinical and bench research domains in biomedicine.

In work funded through the Institute of Museum and Library Science (IMLS),[27] Cornell University, in partnership with Washington University in St. Louis, is developing the Datastar open-source semantic platform as an extension to VIVO. Datastar aims to enable research data sharing and discovery services and to expose metadata about research datasets as linked data. The Datastar ontology will be designed to complement and be compatible with the VIVO ontology in order to support selective interchange and integration of information between VIVO and Datastar RDF stores.

2.7.1   INTERNATIONAL PARTNERSHIPS

In the past year, the VIVO project has also established two significant partnerships aimed at building international consensus around the modeling of research activities and outputs in a technology-

---

[24]http://bioportal.bioontology.org/
[25]https://www.eagle-i.net
[26]http://ctsaconnect.org
[27]http://www.imls.gov/

neutral way. In November 2011, VIVO and euroCRIS[28] announced a joint cooperation effort focusing on interoperation and convergence of the VIVO ontology and CERIF[29] models of research information with a goal of providing homogeneous access over heterogeneous research information systems. The euroCRIS Linked Open Data Task Group, led by Dr. Miguel-Angel Sicilia of the University of Alcalá, has set goals for the calendar year 2012, which include developing use cases for providing linked data (both open and closed) from current research information systems (CRIS) and creating an official CERIF-VIVO mapping as a collaboration between the Food and Agriculture Organization of the United Nations AGRIS[30] initiative and the European Union project VOA3R.[31]

In April 2012, the VIVO project also announced a partnership with the Consortia Advancing Standards in Research Administration Information (CASRAI),[32] a non-profit standards-development organization with headquarters in Canada and chapters in the UK and U.S. CASRAI is also a euroCRIS strategic partner, and by bringing together North American and European partners is playing a leadership role in "standardizing the data that researchers, their institutions and funding agencies must produce, store, exchange and process throughout the life-cycle of research activity."[33] These efforts aim to improve the quality of data about research, streamline the creation and exchange of that data, and simplify the assessment of the impacts of research in societal as well as academic dimensions.

A third set of discussions has involved the Plataforma Lattes,[34] a database of curricula vitae and institutions maintained by the Ministério da Ciência e Tecnologia of Brazil. Lattes has been explored as a model for efforts in the U.S. to develop requirements for a Science Experts Network Curriculum Vitae;[35] the VIVO ontology has provided one platform for discussion of alignment between the Lattes XML data model and a potential future RDF-based research data exchange standard in the U.S.

2.7.2 FUTURE DIRECTIONS

The legacy of the VIVO project will ultimately depend more on the partnerships and principles of open data interchange it fosters than on any specific features of the VIVO ontology or application, which will both continue to evolve in response to new requirements, technologies, and social practices. The data produced will have a life well beyond the home institution by bringing authoritative information into circulation where it can be searched, browsed, analyzed, visualized, and cross-referenced in the rapidly evolving research discovery and networking infrastructure in the U.S. and internationally.

[28] http://www.eurocris.org
[29] http://www.eurocris.org/Index.php?page=CERIFreleases&t=1
[30] http://agris.fao.org/
[31] http://voa3r.eu/, http://www.ieru.org/voa3r/documents/other/PRESS_RELEASE_VOA3R_march2012.pdf
[32] http://casrai.org
[33] http://casrai.org/about
[34] http://lattes.cnpq.br/english/index.htm
[35] http://rbm.nih.gov/profile_project.htm

The ontology team is developing longer-term goals for VIVO that recognize the rapid expansion of linked open data and with it the amount of outdated and duplicate information on the web. Future work will target improved linkages between VIVO and name authorities for people and organizations offering permanent URIs such as ORCID,[36] ISNI,[37] and VIAF;[38] similar authority services will be equally important to permit alignment of data around journal names, events, geographic locations, and terminology, where organizations such as the Library of Congress[39] and the British Library[40] have taken important initial steps. In discussing alternative approaches for reconciling different sources of information about the same author, Geoffrey Bilder argues the importance of retaining the originally submitted information without merging into a single record, leveraging "sameAs" assertions for unified presentation while not in fact losing provenance information that may become important as additional sources of information emerge over time.[41] It may be equally important to support assertions that information is *not* true, as supported by the W3C OWL 2 Web Ontology Language.[42]

Semantic Web standards and the Linked Open Data movement herald an era of unprecedented data exchange. While there will surely be some challenges with the quality and coherence of the data that emerges, VIVO offers the opportunity to participate and a path to improve both the structure and reliability of data about research and researchers.

ACKNOWLEDGMENTS

We would like to thank the extended ontology team, particularly Paul Albert from Weill Cornell Medical College and Michaeleen Trimarchi from the Scripps Research Institute, for their valuable participation in the ontology team phone conferences. Melissa Haendel and Carlo Torniai from the Oregon Health and Science University Ontology Development Group have provided valuable advice and guidance on ontology best practices throughout the VIVO project and in ongoing collaborations. We also wish to thank Paula Markes for her early and effective organizational efforts in forming the VIVO ontology team and community. ShanShan Chen, Bing He, and Ahmed Mansoor also contributed significantly to VIVO ontology development, documentation, and publications during their graduate studies at Indiana University.

REFERENCES

[1] Bilder, Geoffrey. 2011. "Disambiguation without De-duplication: Modeling Authority and Trust in the ORCID System." March 16. http://about.orcid.org/content/disambiguation-

---

[36] http://about.orcid.org/
[37] http://www.isni.org/
[38] http://viaf.org/
[39] http://lcsubjects.org/
[40] http://www.bl.uk/bibliographic/datafree.html
[41] http://about.orcid.org/content/disambiguation-without-de-duplication-modeling-authority-and-trust-orcid-system
[42] http://www.w3.org/TR/2009/WD-owl2-new-features-20090611

without-de-duplication-model ing-authority-and-trust-orcid-system

[2] Courtot, Mélanie, Frank Gibson, Allyson Lister, James Malone, Daniel Schober, Ryan Brinkman, and Alan Ruttenberg. 2009. "MIREOT: the Minimum Information to Reference an External Ontology Term." Nature Precedings. DOI: 10.1038/npre.2009.3576.1

[3] Dovland, Johan, Einar Broch Johnsen, Olaf Owe, and Martin Steffen. 2009. "Incremental Reasoning for Multiple Inheritance." In *Proceedings of the 7th International Conference on Integrated Formal Methods (iFM'09), Düsseldorf, Germany, February 16-19.* Berlin: Springer-Verlag. DOI: 10.1007/978-3-642-00255-7_15

[4] Noy, Natalya F., Ramanthan Guha, and Mark A. Musen. 2005. "User Ratings of Ontologies: Who Will Rate the Raters?" In *Proceedings of the American Association for Artificial Intelligence 2005 Spring Symposium on Knowledge Collection from Volunteer Contributors, Stanford, CA.*

[5] Raymond, Eric S. 2001. *The Cathedral and the Bazaar: Musings on Linux and Open Source by an Accidental Revolutionary.* Sebastopol, CA: O'Reilly Media, Inc.

[6] Rodriguez, Marko, Johan Bollen, Herbert Van de Sompel. 2007. "A Practical Ontology for the Large-Scale Modeling of Scholarly Artifacts and their Usage." In *Proceedings of the IEEE/ACM Joint Conference on Digital Libraries (JCDL'07)*, pp. 278-287. DOI: 10.1145/1255175.1255229

[7] Smith, Barry, Michael Ashburner, Cornelius Rosse, Jonathan Bard, William Bug, Werner Ceusters, Louis J. Goldberg, Karen Eilbeck, Amelia Ireland, Christopher J. Mungall, The OBI Consortium, Neocles Leontis, Philippe Rocca-Serra, Alan Ruttenberg, Susanna-Assunta Sansone, Richard H Scheuermann, Nigam Shah, Patricia L Whetzel, and Suzanna Lewis. 2007. "The OBO Foundry: Coordinated Evolution of Ontologies to Support Biomedical Data Integration." *Nature Biotechnology* 25 (11): 1251-1255. DOI: 10.1038/nbt1346

CHAPTER 3

Implementing VIVO and Filling It with Life

Valrie Davis, *University of Florida*
Kristi L. Holmes, *Washington University*
Brian J. Lowe, *Cornell University*
Leslie McIntosh, *Washington University*
Liz Tomich and Alex Viggio, *University of Colorado Boulder*

Abstract

Research and planning are as crucial to the implementation process as proper installation of software and strategic ingest of data. Each institution has its own diverse landscape, with regulations, concerns, and needs that affect how VIVO is implemented. Therefore, understanding your institution's unique context is paramount to VIVO's success. The next chapter will assist you with understanding the steps of the implementation journey: from planning and relationship building, through data identification and negotiation, to data ingest and outreach to your community.

Keywords

data, implementation, outreach, project planning, stakeholders

3.1 PREPARATION FOR IMPLEMENTATION

You have made the exciting leap to implement VIVO at your organization, but you are unsure of the implementation steps or the preparation needed to begin. The implementation and preparation steps may vary from institution to institution, but there are several important elements needed for any implementation site.

The importance of documentation and communication cannot be overemphasized, despite the time-consuming nature of both. There are a number of different planning documents you can utilize to communicate effectively. First, the initial project plan is a document outlining a bird's-eye view of the project. This will help you understand your institutional context and is a useful tool for communicating the project to the implementation team. Another document, the project one-pager, outlines the project to people outside the team. Additional types of documents you may create include: sub-project charters and timelines, weekly task reports, and data-acquisition requirements. Covering all the different types of planning documents could fill its own chapter; therefore, in this chapter we limit our discussion to the initial project plan and project one-pager as both are vital to the implementation process.

3.1.1   CREATE YOUR PROJECT PLAN

While project plans may vary greatly in structure and presentation, they incorporate several common types of information. The project plan lays out the overall approach for work to be completed and should serve as the primary point of reference for project goals and objectives, scope, organization, estimates, work plan, and budget. The project plan also serves as a contract between the Project Team and the Project Sponsors, stating what will be delivered according to the budget, time constraints, risks, resources, and standards agreed upon for the project, and should include goals, objectives, risks, and assumptions.

3.1.2   CREATE YOUR ONE-PAGER

After your organization completes the initial planning steps and makes a decision to implement VIVO, you may find it quite helpful to construct a document known as a "one-pager" (see Table 3.1), which will outline the purpose of VIVO for people outside the project team. This document accurately and efficiently conveys information about the local VIVO and should summarize information such as: background and rationale, objectives, project sponsors and team composition, impact, and implementation timeline. This information is critical to communicate basic information about the project to the stakeholders.

The one-pager should also include a brief list or discussion of the challenges VIVO faces within the institution as well as information on what VIVO is, why it is important, and where additional information can be obtained. Detailed information is helpful to include and can assist the local team in defining and describing efforts.

3.2   THE IMPORTANCE OF STAKEHOLDERS

Different groups will be interested in how VIVO works and in understanding the value VIVO has for their specific user group within an organization. By leveraging knowledge of what drives a particular institution, it can be easier and more straightforward to garner greater buy-in within the local community. It is critical to note that it is very difficult for a project team to be aware of all institutes, divisions, colleges, and organizational units within an institution. Stakeholders can serve as advocates and cheerleaders for VIVO, addressing issues in a clear, straightforward manner and providing value to VIVO by offering different perspectives, motivating change in organizational workflows, and smoothing the way to data acquisition, adoption, and evaluation activities around the local VIVO effort.

Table 3.1: Project one-pager *(Continues.)*
VIVO Implementation Project June 1, 2012

Rationale:

The complexity and scale of the university, its scholarship, and its research environment are often bewildering to students, faculty, and staff. A simple tool is needed for discovering what work has been done, who has done it, what work is in progress, and by whom. Research and scholarship discovery will enhance our ability to form collaborations, identify synergies, and avoid redundancy. Joining a national network of research and scholarship discovery opens the university to greater collaboration with new research and scholarship partners across the nation and the world.

Goal:

Design and implement an open, Semantic Web research and scholarship-discovery tool using the VIVO software.

Objectives:
1. Implement VIVO as an information service of the library
2. Identify and implement reproducible automated data ingest
3. Promote through an outreach program to students, faculty, and staff
4. Establish data and service governance for the service
5. Establish operational requirements and implement ongoing services

Table 3.1: *(Continued.)* Project one-pager *(Continues.)*

Project Sponsors and Team:

Sponsors include: VIVO Principal Investigator, Dean of the Libraries, and the Project lead. Team members also include center and library staff members, and team members of the Clinical and Translational Informatics Program.

Impact:

Research and scholarship can benefit from direct, simple, integrated access to information regarding people, grants, concepts, events, locations, papers, datasets, and research resources. VIVO greatly simplifies discovery of research and scholarship, providing access to all.

Faculty and staff activities and accomplishments will be publicly accessible, substantially improving the visibility of university research and scholarship.

Data can be repurposed for websites and reporting, significantly lowering the maintenance required to maintain websites and produce reports, while improving data quality.

Participation in the national network provides substantial new opportunities for identifying research opportunities and collaborators, as well as for measuring changes in collaborations over time.

Table 3.1: *(Continued.)* Project one-pager
Timeline: VIVO will be in production by \<date\>. Reproducible automated ingest procedures will be completed by \<date\>. The project will be
completed in \<date\>.

The input and support the technical and outreach team members receive from stakeholders and champions of the project is perhaps the most important component of a local VIVO instance. Partners can help propel the project forward and provide support; therefore, it is critical to identify key stakeholders in the initial stages of the project.

3.2.1   IDENTIFYING STAKEHOLDERS

Stakeholders come from a variety of groups on campus and help communicate the value of VIVO by advancing the work of the local VIVO team in areas related to data acquisition, adoption, and evaluation. They come from a variety of populations, such as administration, faculty, information technology personnel, support staff, and students. Identifying motivating factors among and across stakeholder groups can make interactions with the stakeholders easier and may significantly improve the VIVO team's successful engagement of these groups.

3.2.2   WHAT MOTIVATES STAKEHOLDERS?

Stakeholders can be motivated by a number of factors and environmental pressures. Understanding these motivations can help the VIVO team better partner with the stakeholders in a productive and efficient manner. Consider the following questions:

- What motivates the stakeholders and what are their requirements and expectations?

- How much do they know about VIVO within the organization?

- What do the local VIVO team and a successfully implemented local VIVO instance have to offer the stakeholder?

- How can each stakeholder help the local VIVO with obstacles and objectives?

3.2.3   ENGAGING STAKEHOLDERS

Once stakeholders have been identified and their motivations are better understood, it is time to engage them in a meaningful manner around the project efforts. Stakeholders should be identified at all levels within an organization, ensuring that the most appropriate people serve as the best advocates for the project.

The local VIVO team should be prepared to show examples of various ways stakeholders can communicate the value of VIVO. It is important to remember that there is no one-size-fits-all recipe for stakeholder engagement. You may find that some of your most significant opportunities from feedback, new ideas, or partnership opportunities can come from what seem like the most unlikely places. A list of stakeholders and some perceived benefits can be found on the VIVO SourceForge community site [1].

3.3   IDENTIFYING SOURCES AND NEGOTIATING DATA ACCESS

Identifying and prioritizing your data sources is another step in the implementation process. Determining where data exist, who manages and controls access to the data, and how to get acquisition approval for the data are vital steps in having a successful implementation of VIVO. Approaching data stewards can be a challenging experience if not carefully planned. View these meetings as an opportunity to learn more about your institutional data and as a way to build strong, lasting relationships.

Understanding the organization of the institution will greatly facilitate acquiring the data and permissions, while prioritizing what data you want in VIVO will facilitate where to direct your energy. Use the fields in VIVO to set priorities. For example, many institutions determine that certain minimal data are necessary for VIVO to demonstrate its value (e.g., name, title, position, photo, education, and publications). For greater adoption, data related to research overview, grants, and awards may be desired. Data decisions will vary with each institution and may hinge on the data availability and quality as well as institutional perceptions about data.

You should then determine the local data sources, their management, and what meetings need to be initiated to obtain the data. Human resource departments typically maintain much of the minimal data needed for VIVO, such as name, title, and education, while research or sponsored program offices will have grants and award data. Realize that these organizations are unlikely to have any mandate to collect or maintain data for public consumption; their databases are designed to serve the business functions of the institution. They may be hesitant to turn over data not only because of privacy issues, but also due to concerns about data errors or readability. You might consider including external data sources such as PubMed and other bibliographic metadata aggregators for scholarly publication information, NIH RePORTER, for information on NIH-funded research activities, or other external sources of data. Some work on person disambiguation may need to be done prior to ingest. Internal data often has an associated local identification number which can be useful for disambiguation purposes.

For internal data sources, it is vital to initiate discussions with the data providers early in the process. It is important to understand what issues and concerns drive the data provider, to better meet their requirements and instill trust. Common concerns by data providers across institutions often relate to privacy and security. A pre-emptive step, then, is to consult with your privacy and security officers to acquire support and understanding of the local regulations that may affect how you use the data before meeting with the data provider.

For your first meeting with a data steward, you should be prepared to demonstrate VIVO using another institutional site, present your project plan, ask for test data, and field many technical and policy questions. If you can, become acquainted with the steward's data before this meeting; this can be accomplished by asking for their database fields and acquainting yourself with how those fields correlate with the VIVO ontology. Many questions can be anticipated by reading VIVO materials or by asking local sources. It has been our experience that privacy concerns will be on the top of the list, particularly when it comes to student information. Data custodians may also inquire:

- how individuals will log into VIVO to edit their profiles;

- if there will be a policy in place to opt-in/opt-out of VIVO;

- which system will be the system of record (where data updates occur);

- how emergency removal of sensitive data may be accomplished;

- how frequently data are refreshed;

- how VIVO will be sustained at the institution.

Meetings with data stewards will be most successful with the correct people in the room. From the VIVO team, this typically requires a person adept in data structures and analysis as well as a person who can navigate the political landscape of the institution. Other necessary attendees will be those who understand the data and who have the authority to provide technical access to the data. Timely follow-up communication will facilitate developing strong relationships, which in turn makes data acquisition easier. Through email notifications, telephone conversations, and visual presentations, the stakeholders should be apprised of new or updated data and new platform features, as well as be advised of any issues that may have arisen with the data.

3.4   FILLING VIVO WITH LIFE

Congratulations! You have made it past the initial planning and data acquisition stages and are ready to begin adding life/data to VIVO. There are a number of ways to add data, including: utilizing the Harvester, the ingest tools, manual entry, or a combination of these approaches. Below, we talk about the general steps for each and when you might choose one method over another.

Once your VIVO installation is up and running, its success depends on a rich supply of data. The exact types of data and their level of detail will depend on your VIVO's particular goals. Your installation may include little more than employment positions and recent publication data for a large researcher population in a major institution. It may instead focus on richly detailed descriptions of research interests, publications, grants, projects and their outcomes, and more for a select set of scientists in a small institute. In any case, VIVO depends on structured data. While paragraphs of narrative text often play a role in presenting VIVO profiles for human users, VIVO's unique strengths become evident with data expressed not as narrative text but in bite-sized chunks linked together in meaningful ways. One of the key ways in which VIVO differs from many other web applications is its ability to store and display unlimited kinds of relationships and connections.

3.4.1   MANUAL EDITING

One way to get data into VIVO is to create it manually. VIVO's web-based editing interface allows users to add, modify, or delete all types of data. Different types of users may be granted access to modify the values for different ontology properties; a student library assistant, for example, may be allowed to modify a label and descriptive text about a department, but not to change the list of research grants associated with that department.

VIVO also allows the members of a host institution to update their own information, either by using their institutional login or their own VIVO user account. For example, a researcher may be able to add a paragraph describing his or her research interests, or enter information about a recent publication or presentation. A VIVO administrator has control over what types of data are editable in this way. Researchers may be prohibited, for example, from modifying job position information or courses taught because such data are populated from an institutional database of record.

Any property in any ontology in the VIVO system may be edited, including those introduced as purely local extensions or customizations. For certain properties in the VIVO core ontology, the software includes special customized editing interfaces to simplify certain editing tasks that would otherwise be complicated or confusing in the default interface. For example, when adding authors to a publication, VIVO offers a drag-and-drop interface for setting the author order rather than asking the user to type the author's position in the author list.

The manual editing interfaces enable self-assertion of data, or manual curation by VIVO staff of information for which there is no institutional database or which is unlikely to change rapidly. Manual editing also allows rapid mockups of example data that would ultimately be populated by other means. Manually entering data for a few key people or organizations enables decision-makers to see how VIVO will look and work when, and can be essential to getting buy-in from administrators or those who control access to the types of institutional data sources on which most VIVO installations heavily depend.

3.4.2   AUTOMATED INGEST

Institutional data sources may take a number of forms. Relational databases are often the primary home for various types of data, but the databases in which data are natively stored and manipulated are often not made directly available for query by other systems such as VIVO. Instead, data may be aggregated into larger databases called data warehouses, where complex table structures are simplified for easier understanding and consumption by external systems. Data may also be exposed through web services, using syntaxes such as XML or JSON, or made available for download as comma-separated-value (CSV) spreadsheet files.

While the details of automated ingest can vary widely depending on the type of data and the way it is represented in the source, the process involves several general steps:

1. **Cleaning the Data**

   Many data sources—especially those populated with hand-entered data—require a cleanup step before being used with VIVO. This may be a simple job of stripping out undesirable special characters or trimming extra whitespace, or a more complex task of identifying and correcting typos, spelling errors, and inconsistent entry of names (e.g., "Entomology Dept." versus "Department of Entomology"). Developing heuristics for such cleanup can be time consuming and usually requires a fair amount of human analysis. Fortunately, there are tools to assist with the task, such as Google Refine [2]. An extension developed by Weill Cornell Medical College integrates Google Refine with VIVO and aids a number of ingest tasks.

2. **Producing RDF Triples**

   All data in VIVO is stored in the form of RDF triples. This means that the database tables, spreadsheets, or XML documents provided by data sources must be broken down into a series of simple statements. Producing triples need not be an inherently difficult task. There are tools, for example, which automatically convert relational data into triple form and make it available for query using the SPARQL query language. A W3C candidate recommendation attempts to standardize this technique [3]. Similar direct conversion to triples is possible with other data sources as well, such as spreadsheets.

Table 3.2:			
First Name	**Last Name**	**Job Title**	**Department**
Charles	Wilson	Prof. Asst.	BIO
Thomas	Morse	Lab Tech. II	BIO
Elizabeth	White	Prof.	CHEM

3. **Mapping to Well-Known Ontologies**

   VIVO is not merely a consumer of data; it is also a publisher. By using Linked Open Data and Semantic Web technologies to represent structured data, VIVO enables the data to be reused for purposes beyond VIVO's own public website. Information about particular people or organizations, for example, may be repackaged for display on an organization's web page or loaded into specialized search software for a particular discipline. Most important, data are

ex:n1 <http://example.org/ontology/firstName> "Charles" .
ex:n1 <http://example.org/ontology/lastName> "Wilson" .
ex:n1 <http://example.org/ontology/jobTitle> "Prof. Asst." .
ex:n1 <http://example.org/ontology/department> "BIO" .
ex:n2 <http://example.org/ontology/firstName> "Thomas" .
ex:n2 <http://example.org/ontology/lastName> "Morse" .
ex:n2 <http://example.org/ontology/jobTitle> "Lab Tech II" .
ex:n2 <http://example.org/ontology/department> "BIO" .
ex:n3 <http://example.org/ontology/firstName> "Elizabeth" .
ex:n3 <http://example.org/ontology/lastName> "White" .
ex:n3 <http://example.org/ontology/jobTitle> "Prof." .
ex:n3 <http://example.org/ontology/department> "CHEM" .

Figure 3.1: Example spreadsheet converted to RDF triples.

available to be linked to from across the web, enabling a globally interconnected network of data to supplement the familiar web of text, images, and video.

This interconnected data is interpreted with the aid of ontologies that not only enumerate the types of items described in the data and their relationships with one another, but also enable us to infer things not directly expressed in the data. This is one of the key differences between data stored semantically in VIVO and data stored in most institutional sources.

Institutional relational databases are typically designed efficiently to support a particular set of business purposes: for example, to administer payroll, meet regulatory requirements, or allocate building spaces for teaching and research. These business purposes can be quite different from the discovery and networking goals of VIVO. One of the primary challenges for data ingest is to take data collected for one purpose and make them useful and understandable not only within the VIVO application but on the broader Semantic Web. This involves mapping between schemas for existing data sources and the ontologies used in VIVO, such as the VIVO Core ontology, which, in turn, extends other well-known ontologies such as Friend of a Friend (FOAF) and the Bibliographic Ontology (BIBO). See Chapter 5 for more details about the ontology.

Figure 3.2 shows some of the data from the above example, rewritten to use VIVO Core and FOAF.

In the process of using ontology classes and properties that are well understood by the VIVO software and beyond, we have also gone beyond simple text strings to add additional structure to the data. Instead of merely associating Dr. Wilson with a job title, we have created a named resource (ex:n6606) that represents the position he holds. (The prefix "ex" is shorthand for the namespace

ex:n1 rdf:type vivo:FacultyMember .

ex:n1 foaf:firstName "Charles" .

ex:n1 foaf:lastName> "Wilson" .

ex:n1 vivo:personInPosition .

ex:n6606 .rdf:type core:FacultyPosition .

ex:n6606 vivo:hrJobTitle "Prof. Asst." .

ex:n6606 vivo:positionInOrganization ex:n521 .

ex:n521 rdf:type vivo:Department .

ex:n521 rdfs:label "Biology" .

Figure 3.2: Example RDF data using VIVO core ontology.

part of the resource's URI.) This position is, in turn, linked to a resource (ex:n521) that represents the biology department. This department resource would be linked to anyone else holding a position there, as well as to a variety of other resources. We have also explicitly identified Dr. Wilson as a faculty member by asserting that he is a member of the vivo:FacultyMember class.

Through inference, VIVO (and other applications) can then supply additional triples that we have not created in our ingest process, as with this triple:

ex:n1 rdf:type vivo:FacultyMember.

the software will infer

ex:n1 rdf:type foaf:FacultyMember.

This automatic use of the FOAF vocabulary helps promote interoperability with other Semantic Web applications that may not understand (or care) about the different types of people in academia, but are interested in identifying who the persons are.

By default, the VIVO application creates URIs for resources that end in random integers as a way to promote URI stability, but ingest processes may choose to create URIs in other ways. The URI for Charles Wilson could be, for example, `http://vivo.example.org/individual/ CharlesWilson`. This strategy may work fine for some types of individuals, but will lead to temptation to change URIs when names of individuals—especially people—inevitably change. A safer bet is to mint URIs that incorporate stable, public institutional identifiers. For example, if Charles Wilson has a public employee ID of "cwil2," it may be reasonable to mint a URI of the form: `http:// vivo.example.org/individual/cwil2`.

One of the features of the Semantic Web is that the same resource may be identified by multiple URIs, and these URIs may be linked together using the special predicate owl:sameAs. So it is quite plausible that CharlesWilson may be identified by all three of the above examples by different ingest processes. Ideally, at some point, software will discover (or assume) that these URIs all represent the same person and make the appropriate owl:sameAs links. These processes of entity resolution and disambiguation are a special challenge in data ingest from sources that do not include reliable unique identifiers, such as publication databases which may identify authors by simple strings of the form "J Smith" or "W Wu." Ingest code must identify the URIs that correspond to these authors, or mint new ones.

3.4.3   TOOLS FOR INGEST

Because all VIVO data uses open Semantic Web standards, and all ingest processes ultimately result in the generation of RDF triples, VIVO data ingest can be performed with any number of tools. These include commercial data transformation platforms with graphical interfaces, open-source tools, and custom-built programs.

The University of Florida's Clinical and Translational Research Informatics Program has developed the VIVO Harvester platform as a general, extensible platform for ingesting data from a number of different types of sources, including relational databases, XML files and web services, and CSV files. Mappings between a number of data sources and the VIVO core ontology are available as XSLT transformations. The Harvester enables workflows to be built that extract data from external sources, convert to RDF, map to the VIVO ontology, and apply various scoring and matching techniques to disambiguate names used in the source data and match them to individuals described semantically in VIVO. These workflows are automated and repeatable on a desired schedule. The Harvester also offers CSV-based ingest functions designed to make initial ingest easier for those still learning the VIVO ontology or those unfamiliar with software development and scripting languages. By downloading a spreadsheet template and populating its cells appropriately, users can upload batches of initial data.

The VIVO application also offers some built-in ingest tools for bulk manipulation of RDF data and transformation from sources such as CSV and XML. With these, transformation from one form to another is accomplished by writing SPARQL CONSTRUCT queries. These queries select certain patterns of RDF triples and rewrite a new set of RDF triples. The manual ingest tools can be useful for one-time ingest tasks that do not need frequent updating, or for becoming familiarized with data sources and developing workflows that might later be automated.

3.4.4   UPDATING DATA

Maintaining current, up-to-date VIVO data depends on reliable, automated ingest processes. Ideally, these processes should be repeated as often as possible to ensure VIVO quickly reflects any changes made to source data. There may, however, be important reasons to schedule VIVO updates at regular intervals: daily, weekly, or monthly. If any ingest process relies on human intervention—either to initiate the process or review its results—it is important to manage expectations by setting a schedule that can be reliably kept. It may be better to state that data from a particular source are updated weekly than to claim a daily update but regularly miss days when a staff member is ill or too busy with other tasks.

It is especially important to consider update schedules if an automated process updates the same type of data that is also edited manually through VIVO's web interface. For example, if users are allowed to manually update a person's position title in VIVO, it is important to have a clear understanding of when such titles will be automatically updated by an automated human resources ingest. It may be important to perform such automated updates on a frequency that allows enough time for any errors in the HR database to be corrected before wiping out local corrections made in VIVO.

It greatly simplifies VIVO data management to prescribe that a given type of data will only be updated from a single source. For example, positions and job titles may be updated only via an automated human resources ingest, teaching data may be updated solely from a course XML feed, and research interests and educational background may be updated only by the researchers themselves, using VIVO's editing interface. Such clear delineation ensures that no data will be erroneously erased or modified by another process. It is, of course, only practical where errors in sources of automated data can be corrected in a timely fashion.

3.5   MAKING VIVO A LOCAL SUCCESS

Now that you have data in VIVO, it is time to identify the support system needed to ensure VIVO is both a strong implementation and widely adopted. Every institution/organization is different, with different resources and established work flows. User support can be provided by liaison librarians, an IT help desk, a designated VIVO team member, and so on. You should find a solution that works best in your environment.

3.5.1   OUTREACH AND MARKETING TO COMMUNITY

One of the most critical components of a successful implementation is a comprehensive outreach strategy to market VIVO to the local community and educate about VIVO's features. Many institutions have found that a library-based model of support and dissemination is ideal. Research libraries typically employ information specialists who provide outreach to departments, programs, centers, and individual researchers. These specialists foster enduring professional relationships throughout the academic community and provide assistance to facilitate research. Libraries also maintain a long and valued role on campus as an impartial support partner in a variety of contexts. Through their subject or departmental liaison roles, library information specialists accomplish the following.

- Bring to the project an understanding of both newsworthy and day-to-day activities and issues of importance that inform data element and design decisions.

- Have developed strong and trusted professional relationships with their research clients, and will be able to use these connections to facilitate tasks performed in relation to this project.

- Bring expertise in the areas of collaboration, e-science, digital initiatives, subject specialties, ontologies, controlled vocabularies, and more.

Outreach activities fall into areas related to marketing, education, and support. These activities can be carried out by library liaisons (as described above) or by other supportive individuals who are able to accurately, confidently, and concisely deliver the message about VIVO to the community. Once the liaisons are identified, they must be trained to speak about VIVO in a competent manner in a variety of settings. These outreach ambassadors should be able to make presentations and field questions about data contained within the VIVO. Also, they should be able to understand the structure of the local VIVO team and route questions and offers of support to the appropriate onsite contact. The outreach liaisons should be able to demonstrate VIVO and its capabilities at departments, institutes, centers, and researcher's offices to inform individuals and provide user feedback to the VIVO team to help inform the local efforts.

Outward-facing training about use of the VIVO platform is important as well. Outreach liaisons can host training workshops about a wide range of topics related to the local VIVO, including finding collaborators, editing a profile, and performing more advanced tasks. Outreach liaisons can contribute toward the marketing aspect of the project by developing marketing materials for the local VIVO effort (e.g., flyers, postcards, posters, and electronic announcements for dissemination via email or for projection on announcement screens around campus). Examples of specific outreach activities used at many VIVO sites can be found at [4].

Another significant activity which will help support the efforts of the local VIVO implementation is the establishment of a local advisory board. The advisory board can provide insight on the environmental landscape within the institution and let the VIVO team know about changes on the horizon. The advisory board can serve as a good sounding board for new features as well as request features which may help with the success of VIVO within an organization. The advisory board should be comprised of representatives from the different stakeholder groups, as well as any thought leaders within the organization who may be appropriate, such as administrators from the Office of Sponsored Projects or the provost's office or a representative from a research institute. When contacting people to serve as advisory board members, you should send a brief email or letter that clearly outlines the goals of the effort and the vision for the VIVO implementation within the organization (i.e., What is VIVO? What are the benefits of this approach? What groups are working on VIVO locally, and what is the larger scope of the effort—including other adoptions and work around VIVO by government and funding agencies?). The correspondence should also include the contact details for a representative from the local team who can answer follow-up questions and provide any necessary information.

3.5.2 VALUE-ADDED SERVICES SHARING DATA

The existence of VIVO on a campus adds momentum to complex data initiatives that benefit from collaboration and shared resources. Data solutions can be expensive to buy or to build, which often leads to important data efforts becoming stalled due to limited human and capital resources. VIVO is often the first broad-spectrum public use of faculty and researcher profiles at an institution. VIVO's web infrastructure provides a distributable, machine-readable description of data that allows for stronger data and smart web application linkages. These smart applications include mashups (see Figure 3.3) providing new services or visualizations, data integrations supporting business intelligence or research, and linked data search engines like vivosearch.org. By breaking data out of traditional database silos, VIVO promotes a network effect at an institution and its affiliates where the value of making linked data available increases with the amount of linked data and smart web applications that are available, not only on the campus, but across the Linked Open Data cloud.

Simplifying the sharing and reuse of faculty and researcher profile data on various campus and department websites is a common example of a value-added service of VIVO. With the growing adoption of Semantic Web-aware content management systems like the Drupal 7 open-source CMS, data sharing between VIVO and campus websites will be supported. OpenSocial VIVO gadgets enable existing websites and social software applications to access linked data. Not only is the VIVO source code open, but the VIVO ontology and RDF interchange model are also open so that VIVO profiles are not confined within a walled garden, encouraging faculty and researchers to invest in these portable linked data profiles.

3.5.3 THEME ELEMENTS AND CUSTOMIZATION OF INTERFACE

The VIVO brand is a required element that is provided to implementers. The VIVO identity guide contains important details about how to easily incorporate the specific VIVO branding elements with an institution's brand, logos, and style requirements. Print and web logos for VIVO are provided for download. Flexible FreeMarker templates and custom page layout and styling allow customization of order, position, and location of desired theme elements. Management of property and class groups allows freedom for an institution to focus on groupings relevant to their culture. For example, the People class group distinguishes classes for faculty, staff, students, emeritus faculty, and more—broken down in much detail. If your VIVO implementation is not including certain groupings such as staff or student profiles, those class groups can be hidden or removed. Flexible options exist in VIVO to allow institutions to customize the look and feel of their website theme, navigation, and site content to better represent the unique culture and attributes of their campus.

Navigation
Custom tab-menu management allows institutions to modify the navigation structure as desired. Basic tabs such as People, Research, and Events are included. Highlighting faculty awards or campus outreach may lead to the addition of tabs for these areas. The lack of data from a reliable or useful source may lead to removing some of the basic tabs that will not be used.

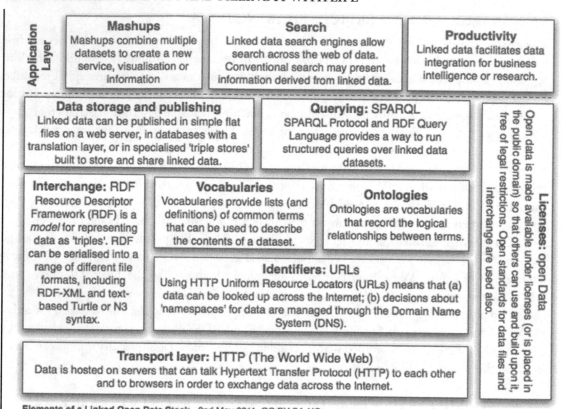

Elements of a Linked Open Data Stack - 2nd May 2011. CC BY-SA-NC
Draft sketch by Tim Davies (@timdavies / tim@practicalparticipation.co.uk) for IKM Working Paper on Linked Open Data for
Development. Comments welcome. Search 'linked open data stack' on http://www.opendataimpacts.net for latest version.

Figure 3.3: Various models and standards that allow for greater exchange of data [5].

Content

Although VIVO provides much underlying data structure through the use of a collection of formal ontologies, the current ontologies may not cover all areas related to academic activity. The opportunity exists for institutions to extend an ontology to better represent their local needs or to contribute a reasonable controlled vocabulary for use by the larger VIVO community. No external ontology tool is needed as the functionality for ontology editing is supported within VIVO.

3.5.4   CONCLUSION AND GENERAL TIPS FOR SUCCESS

Populating VIVO with data is one of the most challenging aspects of implementation, but choosing to implement leads to opportunities for new relationships, prospects, and conversations at the local, national, and international levels. As you begin to outline your steps and embark on your journey, keep in mind the general tips for success that other implementation sites have found helpful. These tips include the following.

- Define achievable milestones, whether those milestones are for an established time period or for a task. Milestones are defined by what works best for your team.

- Show visible interim results within each milestone. If a milestone is one week, identify what tasks each team member will be working on during that week.

- Be realistic and don't go it alone. Milestones can be difficult to achieve, so be sure to accurately assess how much time you need to complete each task. If a roadblock appears, don't hesitate to discuss it as a team and reassign work as needed (if time- or skill-related). Call on your administrators to provide you with support.

- Break through roadblocks. Don't let the small details slow you down. Be sure to know who makes the decisions when disagreement occurs.

- Encourage open, positive communication among your VIVO team members.

- Celebrate victories and work together to understand challenges.

- Take time to do it right the first time. Be sure to understand the data, outline the process, and work the plan correctly.

ACKNOWLEDGMENTS

The content of this chapter represents the work of many people. The authors would like to thank their colleagues at all VIVO implementation sites for their interesting questions, helpful discussions, and great ideas.

REFERENCES

[1] "VIVO Benefits." 2012. SourceForge. Last modified March 4. `http://sourceforge.net/apps/mediawiki/vivo/index.php?title=VIVO_Benefits`. 40

[2] "Google Refine—A Power Tool for Working with Messy Data." 2012. Google. `http://code.google.com/p/google-refine/`. 43

[3] Arenas, Marcelo, Alexandre Bertails, Eric Prud'hommeaux, and Juan Sequeda, eds. 2012. *A Direct Mapping of Relational Data to RDF*. The World Wide Web Consortium (W3C). Last modified February 23. `http://www.w3.org/TR/2012/CR-rdb-direct-mapping-20120223/`. 43

[4] "Local Outreach Activities." 2011. SourceForge. Last modified June 22. `http://sourceforge.net/apps/mediawiki/vivo/index.php?title=Local_Adoption`. 48

[5] Davies, Tim. 2012. "Elements of a Linked Open Data Stack." *Tim's Blog*. `http://www.timdavies.org.uk/wp-content/uploads/IKMLI-LOD-Stack-Draft-Diagram.png`. 50

CHAPTER 4

Case Study: University of Colorado Boulder

Liz Tomich and Alex Viggio, *University of Colorado Boulder*

Abstract

The University of Colorado Boulder launched VIVO in April 2011. We began our serious exploration of VIVO in the fall of 2010 with a proof of concept built on a VIVO 1.1 Virtual Appliance running on a developer laptop. We then began the implementation project in January 2011 and deployed a production instance in April 2011. CU–Boulder differs from the other early-adopter VIVO institutions as it is not part of the CTSA effort and was not part of the original pilot project funded by the NIH. The effort to implement VIVO CU–Boulder, although collaborative across the campus, has been initiated and developed through the Office of Faculty Affairs rather than through typical routes involving library IT efforts or informatics groups housed in medical school environments.

The structure of the chapter is as follows. Section 4.1 describes how we selected VIVO. Section 4.2 provides relevant background on the University of Colorado Boulder. Section 4.3 describes our implementation strategy. In Section 4.4 an overview of our data sources is provided. Section 4.5 explores the technical concerns we encountered. Section 4.6 reviews lessons learned and Section 4.7 describes the value of VIVO to our campus. Finally, Section 4.8 summarizes how we have contributed back to the VIVO community.

Keywords

faculty data, VIVO community, open source, linked open data

4.1 HOW CU–BOULDER CHOSE VIVO

Administrators at CU–Boulder have always been interested in better ways to comprehensively identify faculty expertise and use that information broadly to support campus efforts. In 2008, the University Communications group was tasked with finding a way to build departmental faculty profiles. They had the infrastructure of the home page and web support but no central data source about faculty. Faculty Affairs had ten years of faculty data collected from the campus faculty reporting tool but no way to share access to the data, which lived in a highly secure Oracle database. The Faculty Research Index (FRI) project was born as the two teams united their efforts. The teams worked for two years to construct a custom search application built on components including the Catalyst Perl web application framework, the Dojo Toolkit JavaScript library, and the Apache CXF web services framework. While these components are all open source like VIVO, they operate at a much lower level of functionality. Although some progress was made, the FRI project did not leverage the network effects of Linked Open Data as FRI was a standard HTML web application built on traditional relational data sources.

In the spring of 2010, the existence of VIVO was brought to our attention by staff in a newly formed Boulder interdisciplinary group, now known as the Biofrontiers Institute. VIVO was exciting because it was open source, focused specifically on faculty data, and provided much functionality right out of the box. Building a search mechanism was no longer needed. VIVO provided the necessary technology and structure. The focus now turned to the data.

Since there was minimal risk involved in testing a VIVO implementation using the VIVO virtual appliance, the FIS team decided to test the VIVO hypothesis. The VIVO virtual appliance ran in Oracle's VirtualBox virtualization product, which was freely available as open source, so there was no licensing cost, and we were able to get everything installed and running by following the VIVO documentation. We then copied over some FIS data in CSV format, and we were able to ingest the data using the VIVO Site Administration menu's built-in tools.

Although VIVO seemed unique and unparalleled, our team looked into a proprietary faculty reporting tool that seemed to offer a solution that could serve as a replacement for the aging FRPA Online reporting tool and provide a data stream to feed VIVO more directly. After some evaluation, it became apparent that this vendor solution would not allow the flexibility that our team required to insure making the most of sharing our faculty data. We decided to keep FRPA Online and put our efforts into a future overhaul of that application rather than risk compromising the relationship with our faculty by implementing a system we could not customize and which, therefore, could not sufficiently meet the needs of our campus.

We fully committed to VIVO and started implementation in January 2011. We did not identify stakeholders or secure additional data sources prior to beginning. Since we already owned or interfaced with many key elements of faculty data through FIS and there was no cost involved, we simply proceeded with creating an initial project plan that would:

1. address the most critical, frequently asked questions about the faculty;

2. utilize the high-quality data we already have collected;

3. follow incremental development and features release based on Agile practices;

4. use source of record data;

5. not establish VIVO as a system of record;

6. serve the faculty; and

7. serve the campus.

The seven items above will be discussed throughout the rest of the chapter. The outcome of our initial VIVO effort at CU–Boulder was a limited release of VIVO restricted to campus users in April 2011. In three months, we were able to launch a new application that provided the campus with easy access to searchable data on faculty research expertise and faculty global connections.

4.2   FACULTY DEMOGRAPHICS AND REPORTING TOOLS AT CU–BOULDER

CU–Boulder is the flagship campus of the University of Colorado system, which also includes campuses in Denver and Colorado Springs as well as the Anschutz Medical Campus. CU–Boulder is one of only 35 U.S. public research universities in the prestigious Association of American Universities (AAU), has more than 3,000 teaching and research faculty, and has more than 30,000 undergraduate and graduate students.[1] The campus embraces an interdisciplinary model which is an ideal fit for implementing VIVO. In addition to the 9 major academic schools and colleges, the campus includes close to 100 research centers and institutes. Research collaboration extends well beyond the campus as there are research partnerships with large centers, federal labs, and joint institutes in the Rocky Mountain West. The strategic plan for the campus includes the desire to promote CU–Boulder as a "…global force for expanding frontiers of knowledge."[2]

4.2.1   FIS APPLICATIONS

A commitment to faculty data has been in place at CU–Boulder since the early 1990s with the development of the Faculty Information System (FIS), an Oracle database created in-house by the campus IT development group. The core data application of FIS interfaces with the enterprise HR system for basic information about faculty, such as name, demographics, and position details. FIS captures supplemental data through critical faculty personnel processes managed by the Office of Faculty Affairs. These additional pieces include tracking the tenure review process, sabbatical eligibility, leave of absence details, retirement contracts, appointment details, degree information, and many other details of interest to campus administrators.

The Faculty Report of Professional Activities (FRPA) Online is a web application that has served as the reporting tool for faculty annual reporting since 1998. FRPA Online's first purpose was to collect information used for the annual merit evaluation process. Basic research interest information, entered as codes from a standardized list of research keywords, was included from the beginning.

FRPA quickly became a centralized source of record for faculty research interests and productivity and a default mechanism for collecting new data from the faculty.

[1]University Communications, *Colorado Discovery & Innovation*, vol. 1(4), Winter 2012, CU-Boulder.
[2]University Communications, *Colorado*, vol. 1(1), Winter 2011, CU-Boulder.

4.2.2   FIS TEAM

The FIS Development Team is a small team made up of two software developers, a product owner, and two support staff providing reporting and data support with occasional support from graduate students. Leveraging the power of VIVO is especially useful for a small development team. See a discussion of team practices in Section 4.5.

4.3   IMPLEMENTATION STRATEGY

This section will discuss the core components of the implementation strategy at CU–Boulder. The primary goal of the FIS team is to gather the best faculty data possible in a centralized place and use it everywhere to fill the extensive data needs of the campus. This centralized collection effort:

1. involves data that are entered once and used many times in multiple applications;

2. reduces the reporting burden on faculty and staff;

3. allows for comprehensive data review;

4. creates data streams that become highly visible resulting in motivation for data cleanup; and

5. brings data initiatives together, allowing beneficial leverage on effort, decisions, and resources.

Though much of the national VIVO effort has developed out of the bioinformatics space due to the initial NIH funding, CU–Boulder is deeply committed to a VIVO instance that represents the Boulder faculty to the broadest range—from scientists to humanists. This commitment, coupled with our strategy of providing auto-populated profiles rather than manually created profiles, will require creative approaches to data elements such as publication data where data sources related to the scientific community are much more centrally organized. In addition, CU–Boulder has developed a clear definition of the minimum requirements for a VIVO profile—faculty must be in active status and must have reported either research interests or international activities. It is not enough for a profile to show name, rank, and department. Emeritus faculty are important to represent, but they will be brought into VIVO in a future phase of development in order to address the specifics of this population of faculty.

The strategy employed at CU–Boulder has been to consider the questions asked most frequently about the faculty and to use the best data available from sources that are considered sources of record. The questions asked most often, and most urgently, relate to knowing research expertise and global connections. Fortunately, FIS has collected this information for a while, and with the improved modules added around research interests, these data elements were available for the initial launch of VIVO. Much of the data of greatest interest was self-reported. With high regard for the VIVO concept of provenance and verifiable, institution-backed data, a curation process has been implemented to insure quality information in the area of research interests submitted by the faculty. Other areas may be identified for inclusion in the VIVO curation process in the future. For now, most research interest elements posted to VIVO are approved by an individual curator in Faculty Affairs prior to posting to VIVO. The curation module was built using Oracle APEX and consists of an elaborate workflow to track research interest submissions. Items not meeting the defined criteria or format are returned to the faculty member for resubmission. In this way, Faculty Affairs can insure the highest-quality representation for the faculty and the campus.

Since the FIS team follows Agile software development practices, it was a natural conclusion that our VIVO would roll out incrementally. Feedback from the campus, and especially from the faculty, has been incredibly useful in guiding the direction of the project to create a valuable product that properly represents the faculty. Our original project plan is mostly still intact, but our priorities have been flexible and adjusted as needed, whether due to policy concerns that created roadblocks requiring software solutions or to the appearance of new, highly motivated partners wishing to collaborate.

The decision to limit the initial VIVO launch to campus users prior to going fully public has proved to be essential to the success of this project. Going public with data that have been historically used for internal evaluation requires clear communication, often through multiple channels, to allow for comprehensive data review by the individuals and consideration of their concerns about how the resulting profile will look.

4.4 SOURCE DATA FROM ENTERPRISE SYSTEMS RATHER THAN BUILDING NEW PROFILES

VIVO profiles at CU–Boulder are populated from source data using VIVO's Data Ingest. At CU–Boulder, VIVO is not the system of record. Depending on the data element, the system of record is FIS, the enterprise HR system, the enterprise student information system (ISIS), or any number of connecting systems that will feed into VIVO CU–Boulder in the future (see Figure 4.1).

Faculty do not have access to enter or edit anything directly in their VIVO profile. All data collection is funneled through the FRPA Online web application. Although initially created to support the annual merit evaluation process, FRPA Online has become the main data collection point for many additional data elements submitted by the faculty, including research interests, international activities, outreach efforts, and profile basics such as preferred name and a profile photo.

Prior to the initial VIVO launch at the Boulder campus, it became apparent that the information on research interest collected through FRPA Online needed to be expanded. The FIS team was able to release a greatly expanded module to coincide with the reporting cycle timelines right before the intended VIVO implementation. Although collected for many years, research interest data was originally limited to the faculty selecting codes from a standardized list of research keywords. This feature was cumbersome and extremely limited by the topics on list. The FIS team added the following data entry enhancements to better support a VIVO instance for the campus:

1. Free-Text Keywords—words or phrases entered by faculty members to describe their research.

2. Web URL—no central repository existed for faculty URLs previously.

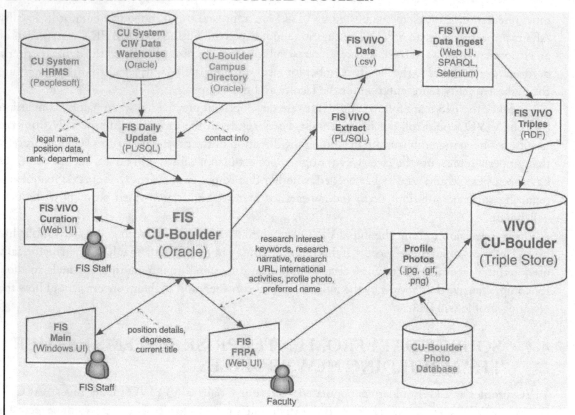

Figure 4.1: Data Ingest for VIVO CU–Boulder.

3. Research Overview—a brief overview of their research.

4. Sensitive Research Overview—to allow faculty to discuss research they did not want posted on a public VIVO profile.

A module to collect information on International Activities was created a few years prior as the campus embraced initiatives around global connections related to student and faculty recruitment as well as increasing diversity and educational offerings. Country data collected from these questions are posted to the VIVO profiles, so discovery may occur, highlighting global connections indicated by the faculty. Improved detail is considered for future development in this area.

CU–Boulder's FRPA went from 12 years of internal use for evaluation and faculty productivity analysis to jumping immediately to public use for some data elements. The initial thought was that the limited information to be shown on VIVO profiles was already public, for the most part, on departmental websites or on faculty vitas posted online. This assumption is mostly true, but the exceptions are critical to understand and have forced our development team to slow the VIVO implementation in order to build new features so the faculty feels protected and properly represented by their profiles. Reputation and recognition are important to faculty careers, and public VIVO profiles must include accurate and timely information to support this.

Faculty feedback has been extremely useful in this regard. For example, we learned that faculty want to completely control what name is on their public profile. Legal name, used for most human resource requirements, was unacceptable in some cases and did not cover the faculty's need to maintain the continuity of their scholarly career. Faculty often identify themselves in citations, course catalogs, or other public venues under the name they are most recognized by rather than what may be their current legal name. We are developing a "preferred name" feature in FRPA to allow for this. VIVO CU–Boulder will remain restricted to campus users only until this feature is in place. We have chosen this route rather than to open up VIVO to any manual entry at this stage.

Academic rank and prestigious titles were an area we knew had to be accurately represented in a VIVO profile. Since the HR job classes serve as the academic rank for most faculty, we have applied a series of algorithms to HR data to properly identify the primary appointment and, therefore, the appropriate rank to post on the profile. Staff in Faculty Affairs who process appointment letters are able to use a feature in FIS to capture additional title details related to special endowed, named, or administrative appointments. The rank posted to VIVO automatically updates based on appointment dates.

At the start, name, rank, research interests, and international activities were identified as the minimum key elements to include in a VIVO profile. Photos were also deemed as crucial to create visual interest in the application. The FIS team secured an initial load of a few hundred faculty photos from the campus photographer and also added a feature to allow faculty to upload their own profile photo from within FRPA Online. Currently, nearly 50% (about 800) of our VIVO profiles display a photo.

The VIVO implementation at CU–Boulder has opened up and accelerated the conversation about faculty data needs. The FIS team continues to address links to source data and create new entry options as needed. By moving forward and building collaboration with other campus data partners, we hope to motivate others to join the initiative and contribute their data or build VIVO features that may be of interest. The channel has been opened specifically with the Data Management Task Force (handling large research data sets), the University Libraries (building a digital repository), and the University Communications group (Drupal web publishing).

4.5   OUR TECHNICAL ENVIRONMENT AND GETTING WORK DONE

Our team's shared values and principles support an incremental process and overlapping practices that have helped our small team to effectively implement VIVO CU–Boulder.

4.5.1   VALUES AND PRINCIPLES

We value **communication, simplicity, feedback**, and **respect** as defined by the Extreme Programming (XP)[3] community. We also highly value **learning,** which prepared us for the new technical knowledge required to best leverage VIVO's Semantic Web toolkit. Our guiding principles are also well defined within XP and include **mutual benefit**, **quality**, **flow**, **economics,** and **accepted responsibility**. Our team values and principles have integrated well with VIVO's open source and open community.

4.5.2   PROCESS

The FIS team adopted Scrum,[4] an Agile software development process. Scrum organizes and focuses our team's effort in time boxes known as Sprints, and it scopes the work to occur in the Sprint using a Sprint Backlog. This backlog is the list of tasks to be performed, which are selected from the larger Product Backlog. We settled on a cadence of two-week Sprints to maximize feedback and flow. Our bi-weekly Sprint Review Meeting provides a retrospective on the work completed as well as areas for improvement. Scrum allows our small team to remain focused by interspersing VIVO Sprints with Sprints focused on legacy FIS systems. This minimizes focus-sapping, out-of-context interruptions—e.g., "Can this new request wait two weeks?"

In 2011, we began to adopt Lean software development[5] and Lean Startup[6] principles and practices. These include **limiting work in progress, Validated Learning,** and **Minimum Viable Product (MVP)**. Lean places valuable attention on the **economics** of software development. Limiting work in progress is critical for a small team to maximize value delivery. As described earlier, our VIVO VirtualBox proof of concept was a useful MVP to validate our hypothesis that VIVO could be substituted for our in-house Faculty Research Index.

4.5.3   PRACTICES

The technical practices applied in our VIVO implementation are also pulled from Lean and XP. These include using a **shared team workspace, user stories, pair programming, test automation,** and **incremental deployment**. These practices have been productive for our small team for maintaining product quality in our ongoing VIVO implementation project and other FIS components.

4.5.4   TOOLS

The open-source tools coming out of the VIVO community lay the foundation for building a stable next-generation FIS, without placing external constraints on our freedom to improve and innovate. In addition to the open-source tools that are bundled or associated with VIVO, like MySQL, Jena, and

---

[3]Kent Beck and Cynthia Andres, *Extreme Programming Explained* (Boston: Addison-Wesley/Pearson Education, 2005).
[4]Ken Schwaber and Mike Beedle, *Agile Software Development with Scrum* (Upper Saddle River, NJ: Prentice Hall, 2001).
[5]Mary Poppendieck and Tom Poppendieck, *Lean Software Development: An Agile Toolkit* (Boston: Addison-Wesley/Pearson Education, 2003).
[6]Eric Ries, *The Lean Startup* (New York: Crown Business, 2011).

Tomcat, additional open-source technologies that have been useful in our implementation project include **Selenium**,[7] **Trac**,[8] and **Git**.[9]

For example, Selenium is a browser automation tool that we previously used to automate acceptance tests. We began using Selenium to automate our VIVO configuration and data ingest with the early VIVO 1.1 **Site Administration** menu's web-based features, as the VIVO Harvester was still evolving in 2010. Our Selenium scripts can configure a new VIVO instance from a fresh VIVO install or one with an empty MySQL database with no human intervention.

4.5.5 IT PARTNERS AND VIVO HOSTING

The FIS team works closely with the Managed Services group in CU–Boulder's Office of Information Technology (OIT) on our VIVO implementation. OIT Managed Services provides managed hosting of our FIS test and production servers. This trusted relationship allows our small team of developers to focus on development and integration tasks and minimizes our team's system administration task load. Since VIVO was built on a common enterprise open-source software stack supported by our existing physical environment, Managed Services was able to contribute expertise in the installation, configuration, securing, and patching of core VIVO components. The FIS production environment consists of 64-bit Dell PowerEdge enterprise-grade servers. The database server is a R710 with 12 cores and 16GB of RAM. The web server is a R610 with 8 cores and 24GB of RAM.

With the goal of modernizing campus-wide web infrastructure, University Communications and OIT selected **Drupal 7**[10] and a stack of primarily open-source technologies for the CU–Boulder WWW NextGen project.[11] The new Drupal-based campus website is scheduled to go live in early 2012, and our team will be provided with a Drupal sandbox to explore VIVO integrations. Our goal is to enable other campus units to reuse VIVO linked data in their Drupal-based websites. Value also exists in understanding the rendering of content, of interest in our VIVO Linked Open Data context, in Drupal 7 websites with **RDFa markup**.[12]

4.5.6 CONTINUOUS IMPROVEMENT

Our team plans to add another full-time software developer in 2012. With this additional resource focused on VIVO implementation and development, we will continue with incremental improvements to our team process and practices, including integrating **Kanban**[13] into our process, more opportunities for **pair programming**, **test-driven development**, and **continuous integration** to support new VIVO development. An emerging topic of future interest is **DevOps**.[14] These are principles, methods, and practices to optimize communication, collaboration, and integration be-

---

[7]http://seleniumhq.org/
[8]http://trac.edgewall.org/
[9]http://git-scm.com/
[10]http://drupal.org/
[11]http://www.colorado.edu/about/wwwnextgen
[12]http://rdfa.info/
[13]David J. Anderson, *Kanban: Successful Evolutionary Change for Your Technology Business* (Sequim, WA: Blue Hole Press, 2010).
[14]http://en.wikipedia.org/wiki/DevOps

tween software developers and IT administrators. As our VIVO implementation grows, this will be an area for close examination to see how our small team can best improve our collaboration with OIT Managed Services.

4.6   CHALLENGES

There were no significant technical challenges related to implementing VIVO at CU–Boulder. Even without using VIVO's Harvester to load important data streams, VIVO CU–Boulder is stable and secure and provides campus users with easy access to useful information. Most challenges for our implementation involved building the campus initiative, solving data quality issues, and addressing faculty concerns about privacy or how they would be represented in a public profile. Data quality issues and the subsequent solutions have already been discussed (see Section 4.4).

Faculty expressed additional concerns about privacy to protect their work from competitors, especially in terms of details about grants or works in progress. Neither of these areas is slated for public posting on our VIVO at this point in time, if ever. This concern does point to the challenge that arises from using a reporting tool historically focused on internal communication for evaluation and selecting portions for public display. Although public data elements are clearly marked at the point of data entry, there is concern that all items reported could be used on the public profile, which creates tension for some faculty. Future plans are to separate the data collection efforts in order to give faculty peace of mind to detail their work for the internal audience.

Building trust with the faculty is an important communication effort which, in this day of increased litigation and complex traditional enterprise IT solutions, is more important than ever and harder to achieve due to the high expectation for IT solutions and the often less-than-ideal IT environments that exist. Our VIVO project has high standards for response and data quality, so we endeavor to listen carefully to our faculty. The flexibility offered by the technology behind VIVO and the open-source VIVO community offer valuable resources for adhering to these standards.

4.7   THE CURRENT AND FUTURE VALUE OF VIVO TO THE CU–BOULDER CAMPUS

VIVO has the potential to become an invaluable resource to the Boulder campus. Faculty are one of the greatest assets to the campus, and VIVO allows quick search and discovery of their expertise and accomplishments. Promoting the campus, building mission-critical initiatives, facilitating funding opportunities, supporting outreach efforts, and creating collaboration among faculty, industry, government agencies, and other academic institutions are all greatly enhanced by the existence of VIVO CU–Boulder. We have engaged with our sister campuses to deliver our Faculty Information System (FIS) suite of applications for their use, which includes the possibility of full and separate instances of VIVO for each campus. This effort, linked by VIVO's functionality, provides cost sharing and builds a common data resource for the entire University of Colorado system.

Building features to enhance our VIVO project has started conversations with many data partners. Data initiatives around publication data are starting to be channeled in one main flow rather than proceeding in smaller, self-contained projects. Not only does this create a cost-sharing possibility, but more importantly, more progress may occur if a larger effort is in place. Leveraging decisions and man-hours builds a more productive solution. Our main partners in this effort include numerous library faculty with special projects, institutional research analysts charged with major reporting objectives at the national level, and the communications group managing departmental websites and the CU home page.

In addition to acquiring new data streams to feed VIVO and other collaborative efforts, data quality, in general, is getting more attention due to the high visibility of a public faculty profile. Faculty have desired a "preferred name" solution for years. Now, because of VIVO, they will gain that alternative over legal name for their career activities. As faculty begin to see their VIVO profiles built through data sources, there will be even more of a concerted effort to update and complete pertinent information.

Although progress is already at hand, the future holds even more promise. VIVO may provide the integral structure for bringing together the multitude of regional research groups in this area to create a larger, more active regional research collaboration in the Rocky Mountain West. Because of its simple, low-cost-yet- effective capabilities, VIVO is an attractive solution to create a shared network among academic institutions, federal labs, government agencies, and private sector researchers so critical to the economic and educational growth of Colorado and its neighbors. The University of Colorado Boulder hopes to be influential in bringing VIVO to these research partners in the interest of creating collaboration, clarifying information, and building room for vital new discoveries.

4.8   CONTRIBUTING TO THE VIVO COMMUNITY

The only way that open source works is through **community participation**. While we initially expected our team's contributions to the VIVO open-source project to be in the form of bug fixes and enhancements, we found that VIVO fit our initial project needs with minimal customizations. Although we have yet to contribute a line of source code, the community feedback we have received indicates that our contributions are valued and may provide ideas for other small implementation teams.

4.8.1   NATIONAL IMPLEMENTATION SUPPORT

The VIVO national implementation team invited us to help with this community support effort. Our primary responsibility in this role has been to coordinate bi-weekly online meetings. These are open to all institutions implementing VIVO, with additional information maintained in the Implementation section[15] of the VIVO wiki on SourceForge.

---

[15]http://sourceforge.net/apps/mediawiki/vivo/index.php?title=Implementation

4.8.2   UNIVERSITY OF COLORADO SEMANTIC WEB INCUBATOR

In 2011, we kicked off local efforts to establish an ongoing Linked Open Data discussion at CU–Boulder. Prior to our VIVO implementation, we had not encountered other Semantic Web or Linked Open Data efforts on campus. As we meet with our campus partners regarding VIVO implementation or other data sharing topics, we are identifying opportunities to collaborate and contribute to a Linked Open Data network effect. We are in discussions with CU's CTSA awardee, the CCTSI,[16] which is implementing the RNS Profiles[17] web application that will present their researcher profiles as VIVO Linked Open Data. VIVO will also be shared with the UCCS campus as part of our FIS implementation.

4.8.3   IMPLEMENTATION WORKSHOP

CU–Boulder will be hosting the VIVO Implementation Fest in 2012. The original two-day format will be expanded to add a half-day introduction to VIVO and the Semantic Web to which we will invite CU–Boulder's regional partners. These include other major institutions of higher learning in Colorado, federal labs, local technology groups, and campus peers. Goals include identifying common national and regional VIVO implementation and Linked Open Data goals, and to open the discussion to the idea of a regional research network.

ACKNOWLEDGMENTS

We would especially like to thank Medha Devare and Jon Corson-Rikert of Cornell University for their initial welcoming of our Boulder team to the VIVO community, as well as a long list of community members who have helped us with our implementation. Special thanks also to Jeff Cox of CU–Boulder for seeing the possibilities, thinking outside the box, and supporting big visions.

---

[16]http://cctsi.ucdenver.edu
[17]http://profiles.catalyst.harvard.edu/

CHAPTER 5

Case Study: Weill Cornell Medical College

Paul J. Albert, Curtis Cole, and the WCMC VIVO Project Team: Dan Dickinson, John Ruffing, Markus Bronniman, Eliza Chan, and Kenneth Lee, *Weill Cornell Medical College*

Abstract

In this chapter, we present the use case for VIVO at our multi-institution Clinical and Translation Science Center (CTSC). We highlight the weakness of our legacy researcher profile system addressed by VIVO; discuss the process of acquiring and mapping data according to the VIVO core ontology; define policies, procedures, and our ingest work-flow; and demonstrate why the software successfully serves our users' needs. Finally, we review our remaining issues, including managing expertise descriptions, implementing self-editing and reporting, and automatically generating a curriculum vitae or biosketch from a VIVO profile.

In 2003, when VIVO was first developed at Cornell University in Ithaca, the faculty at the Manhattan-based Weill Cornell Medical College (WCMC) were invited to participate. At that time, WCMC had already developed its own faculty profiling system that was independent from VIVO. Therefore, WCMC exported data from its own systems and sent a data extract to the main campus containing names, departmental appointments, and link-back URLs. This data extract was transformed into a VIVO-friendly format and loaded into the main campus's VIVO instance.

After the redevelopment of VIVO as a Semantic Web application, the potential utility for VIVO as a primary system at WCMC changed considerably. In October 2009, WCMC joined six partner institutions that were awarded a grant to implement and further develop VIVO. Our work on the VIVO project has been a catalyst for resolving questions about conflicting data, improving the quality and structure of data from our sources, and sharing under-publicized data with local researchers and administrators, as well as independent indices such as Google Scholar that store researcher metadata.

Keywords

researcher metadata, VIVO community, publications, multi-institutional environment, automated ingest, Google Refine+VIVO

5.1   MULTI-INSTITUTIONAL ENVIRONMENT

Although VIVO can serve a variety of purposes, the goal of facilitating collaboration is particularly relevant in our multi-institutional environment. Despite our small geographic footprint on the Upper East Side of Manhattan, due to the complexity of our medical center, one could spend years at the College only to realize that his/her ideal collaborator is only the next building over. This is even more so since 2007 when WCMC formed a multi-institutional collaboration to compete for the National Institutes of Health (NIH) Clinical and Translational Science Award, UL1 RR 024996.

Based in New York City at Weill Cornell Medical College, the Clinical and Translational Science Center (CTSC) is a consortium that serves a number of local partner institutions, including: Weill Cornell Medical School, Cornell University at Ithaca, Hunter College, Memorial Sloan-Kettering Cancer Center (MSKCC), the Hospital for Special Surgery (HSS), and New York Presbyterian Hospital/Weill Cornell Medical Center (NYP). The mission of the CTSC is to translate discoveries at the bench to the bedside and to the community. The goals that align closely with those of VIVO include the identification of collaborators, funding opportunities, and publishing support. We are particularly eager to use VIVO as a tool to match basic scientists with clinicians studying similar topics. VIVO itself has the promise of meeting many of these end-user needs within the CTSC and with our other national and international affiliates.

5.2   LEGACY RESEARCHER PROFILING SYSTEM

Initially developed in the late 1990s, the WCMC faculty profiling system has three "personalities" to serve the different needs of faculty with the typical academic medical center tri-partite clinical, research, and educational mission. The first "personality" was the Physician Organization Profile System (POPS). POPS was designed to deliver clinical profiles of the faculty to patients and referring physicians. Second, the Research Profile System (RPS) personality was added to deliver different, but overlapping content that describes faculty research, including publications and different descriptions detailing their scientific work. The third "personality," called HoPS, extended the clinical profiles to faculty who attend at New York Presbyterian Hospital, one of WCMC's affiliated teaching hospitals. All three profile system "personalities" were built on top of the Java-based content management system Fatwire,[1] although both POPS and RPS existed earlier as standalone WebObjects applications.

RPS has served to aggregate authoritative metadata about researchers with a Weill Cornell faculty appointment. RPS offers approximately twelve major fields of metadata about researchers. Several critical fields, including name, appointment, department, and education, are automatically

---

[1]Now part of Oracle:http://www.oracle.com/us/corporate/acquisitions/fatwire/index.html

ingested from systems of record such as Human Resources and Faculty Affairs. For clinical faculty, additional data is also imported from authoritative systems of records, such as the Managed Care database for inclusion of accepted insurance plans.

A variety of more customized data is manually entered, such as biography, research overview, publications, honors and awards, committee service, global health participation (e.g., statement, countries of focus, languages spoken, percent time dedicated to global health, and areas of interest). Separate biographies and statements are kept for research, global health, and clinical profiles, though common fields are shared.

One of the key features of POPS is the ability to manage medical expertise and specialty terms in profiles. These terms are tied to SNOMED[2] and the Intelligent Medical Objects (IMO)[3] smart search technology, which allows patients and referring physicians to search for doctors based on disease or procedural expertise. This is discussed in more detail below.

5.2.1   ADVANTAGES OF THE LEGACY SYSTEM AS COMPARED WITH VIVO

The primary advantages of the WCMC legacy system over VIVO relate to editing functions and to the non-research-oriented features, especially those relating to clinical activities. As we learned from the initial POPS deployment, a significant majority of faculty members fail to consistently maintain their profiles. To address this problem, VIVO emphasizes automated population of profiles from systems of record. However, bulk editing by delegates and WYSIWYG self-editing are critical features of our legacy system. This allows departmental administrators and other proxies to maintain profiles for busy faculty without their active participation.

Given the original purpose of our legacy system to support patient and referring physician searches, it is not surprising that it contains strengths not found in the research-oriented VIVO system. Expertise searching is the key differentiating feature and one that is very hard to implement. Like many clinical profiling systems, our initial effort was simply a free-text, keyword-based search, much like that found in VIVO. Faculty (or their proxies) can add keywords describing their expertise that can then be searched.

The problem with this approach is that it is not conceptual. An expert in HIV should be found when searching with the word "AIDS." This requires a structured terminology with a hierarchy or ontology. While one of VIVO's strengths is that the data structure itself is based on a shared ontology, implementation sites have yet to represent expertise using a controlled vocabulary. The unique needs of VIVO implementation sites make it challenging to select a particular terminology applicable across all disciplines. A VIVO implementation at an engineering school needs very different terms than "medicine," "architecture," or "the humanities."

Towards the end of this chapter, we describe our emerging approach to expertise searching in the medical domain. The complexity of our particular field illustrates how difficult it will be to

[2]http://www.ihtsdo.org/snomed-ct/
[3]http://www.e-imo.com/

engineer a solution across domains, but also why the fundamental semantic architecture of VIVO holds promise for a future solution.

5.2.2   SHORTCOMINGS OF THE LEGACY SYSTEM AS COMPARED WITH VIVO

The legacy system has several shortcomings, some of which are addressed by VIVO and some of which are not but can guide future development. The major problems relate to self-editing and data accuracy, research expertise, and lack of multi-entity support (e.g., WCMC and our CTSC partner institutions).

While POPS/RPS offers sophisticated options for self-editing, faculty are rarely motivated to keep their data up to date. In our legacy system, publications must be manually updated. As of this writing, one of the College's most prolific researchers has not updated his list of publications since 2006, meaning that at least 67 journal articles are not listed on his profile. VIVO resolves this problem with automated publication ingest.

The ingest capabilities from authoritative systems is possibly the most powerful way to deal with the problem of inaccurate and missing data. But even when our legacy system used "authoritative" data, we quickly learned that we needed to offer the ability to correct or tweak these data. Self-editing is therefore a necessary feature and a trouble-prone problem. Another problem with self-editing that may repeat in VIVO is abuse of text area fields. One user's research overview consists of 7,300 words and is rendered nearly unreadable by the complete absence of visible line breaks. A WYSIWYG editor helps, but we have been consistently surprised by researchers' general lack of concern for the formatting of the final output. Our homegrown system contributes to this problem by offering few controls to help standardize typography and information structure. *There is an inherent tension between user control and content standards that all systems like this will face.*

Finally, our legacy system lacks any straightforward method for data exchange with other intra-institutional and extra-institutional systems. By contrast, VIVO easily allows researchers' RDF triples to be copied into any other Semantic Web system that can host such data.

5.3   PREPARATION FOR INGEST

5.3.1   SETTING UP THE ENVIRONMENT

The first step in bringing up VIVO is setting up the physical environment. We initially decided to create separate development and production environments on a Sun SPARC Enterprise T5240 server from Oracle. The T5240 is a virtual machine with 32 virtual CPUs (4 cores) and has 16 GB of RAM. Unfortunately, latency issues required that we restart the server almost daily for reasons not yet determined. We are now in the process of using a Linux-based server, which has superior performance to our configuration on the Oracle server.

5.3.2   NEGOTIATING WITH SYSTEM OWNERS AND OPERATORS

The most complicated step in setting up VIVO is acquiring authoritative data, including names and definitions, from the various system owners and relevant stakeholders. Below we describe some of the problems we encountered. From discussions with colleagues at other institutions, these seem illustrative of the kinds of challenges that are typical of VIVO implementations.

Our HR system was being upgraded at the time of ingest, so much of the data, such as organizational unit definitions, was still in flux. In other words, the authoritative system was not yet authoritative.

Many of the source systems have no notion of semantic class or property. Data hierarchies are not rigorously defined but pragmatically defined for the purpose of the source system's original setup. For example, in our institutional definitions, one would be wrong to assume that the Stroke Center is a "center"—it's a division! The same problem would repeat with the organizational structures in most of the systems we touched, which often were not internally consistent.

Each informational system views the world through a particular lens. In one part of our agreements and grants database, Coeus,[4] organizational units are defined by an external definition that tends to align with funding agency definitions. In the case of the research accounting system, the worldview is foremost about which internal organizational unit pays which person. As a result, these are all broken out in considerable detail. Similarly, clinical profiles used for marketing physician services are most concerned with where a doctor practices, so practice location is conceptually more important than department or division. Zeroing in on the purpose of a system speaks to how the data is structured. A public and semantic system like VIVO is either another worldview or must have a very robust ontology to account for the semantic diversity of the source systems.

Some system owners were very protective of their data, while others shared freely. In hindsight, we have learned how critical it is to find the correct person to whom you can explain VIVO to get their endorsement. We have learned through experience that our researchers appreciate not being bombarded with questions and updates for their profiles. So emphasizing how VIVO can save them from having to enter the data manually and potentially provide valuable reports is key to getting needed support.

We also learned that by creating local extensions to the VIVO ontology for data inside VIVO, we could help key Coeus users later when they want to use VIVO for reporting. For example, research leadership has recognized the value in faculty profiles that include research activity, publications, and teaching for central productivity reporting and statistical summation that require adding custom classes for different kinds of research agreements.

5.3.3   FINDING AND CLEANING AUTHORITATIVE DATA

In addition to the direct benefits of implementing VIVO, we also looked at participation in the project as a rare opportunity to resolve some of the persistent questions about a dozen different systems of record. Even if the VIVO software failed to fulfill its promise, our commitment to this project could

---

[4]http://osp.mit.edu/coeus/

be a success if we could define which systems recorded institutional data authoritatively and clarify the provenance of metadata about researchers. After go-live, this value would perpetuate as VIVO would provide individual faculty unprecedented transparency into "authoritative" data about them and offer a chance to fix inaccuracies.

As discussed above, some variation in accurate data is natural due to the different views of related information used for different purposes. People have multiple titles: official titles conferred by human resources, academic titles denoting official rank, clinical titles reflecting hospital appointments, and administrative titles that may be unofficial but often most accurately capture someone's job. Even names have this kind of variability as some people decline to be known by their legal name or emphasize a second name rather than the first. Even so, one way or another, we need to identify a single authority for ambiguous data while also figuring out the best public representation of that data.

Surprisingly, the names of organizational units were also more heterogeneous than we expected. For example, when presenting to clinical populations, we refer to one department as "Neurology" while the academic website for the department calls itself "Neurology and Neuroscience." This is another example of the mismatch between clinical and scientific or public and private personalities of an academic medical center.

Once an authoritative source was identified, the next challenge was determining a feedback loop so that errors could be sent back to the source for correction. This is critical for the credibility and acceptance of VIVO. For now, this is mostly phone calls and emails, but we hope for more automation in the future.

5.3.4   MANUAL ENTRY VS. AUTOMATED INGEST

With VIVO, we have committed ourselves to automating data ingest whenever possible. In the enthusiasm of implementing a new system like VIVO, it is tempting to "just get it done" and allocate the resources necessary to populate it manually. This option was seriously considered in the case of our authors' publications. We calculated that entering all the institutions' publications in manual, or at least semi-automated fashion, might cost approximately $50,000. But what then? According to Scopus, Weill-affiliated authors publish about six new journal articles a day. The number of publications to be represented may nearly double when one considers new hires.

Another liability of manually entering data in a semantic application like VIVO is that it is very easy to enter multiple instances of the same object. As a result, one might see a co-author visualization with three identical names. This is further complicated when names are not identical, as is often the case with publication metadata. For example, do the following co-authors refer to the same people: "Hamlim-Cook, Pamela" and "Hamlin-Cook, P.?" What about "Zhao, K. Y." and "Zhao, Kesheng?"

Also, there is the problem of expired content. For example, VIVO can track committee participation, which is important for faculty promotions and productivity tracking. But many committees convene for an undefined length of time. Some committee members may delete this bullet from their profile while others might leave it on to keep track of historical work. This inconsistency could cause a visitor to lose faith in the currency of the data in VIVO.

Thus, far, we have only manually entered data in VIVO in the cases of names of organizational units (e.g., departments, divisions, affiliates, and core research facilities), which tend to change rarely.

5.4 POLICIES AND PROCEDURES

It may be possible to define all of an institution's policies and procedures for VIVO from Day 1, but our thinking about VIVO has certainly evolved with experience. For example, we now focus more on obtaining authoritative data at the expense of less data and greater end-user freedom. Examples of some of our lessons learned regarding policy and procedure follow.

5.4.1 INACCURATE DATA

When soliciting support for VIVO from system owners, we contended that publishing their data in VIVO—whether through end-user feedback or our determinations of where data was authoritative—could ultimately improve source systems' accuracy.

As our production version of VIVO has gained visibility, we received feedback from researchers. Thus, far, there have been only two reasons why data in our VIVO is wrong:

- *The data is inaccurate in the system source itself.* We have received a couple of complaints, for example, that the name of the school a post-doctoral researcher attended was inaccurate. We promptly forwarded this feedback to the owner of the source database, and this issue was resolved in a subsequent ingest. While this is potentially a great feature of VIVO in that it helps correct institutional data by making it visible, the methods for correcting data remain a serious and complex problem. For frustrated users, we cannot guarantee that all the source data will be corrected or corrected in a timely fashion. System owners may not feel these errors matter or may lack the resources to make timely corrections.

- *The data is accurate, but it is mismapped or misloaded.* For instance, a faculty member's PsyD degree was not appearing because the credential was not listed among the individuals in the class core:Academic Degree. In this case, we made a temporary fix until an update to the ontology in version 1.3, which resolved this issue.

It has not yet been necessary to make emergency changes manually in the time before the next ingest, but that is always an option.

5.4.2 SENSITIVE DATA

Even when data is accurate, everyone may not agree on what should and should not be published. For example, there is controversy about how to handle requests to remove email addresses from public profiles. Similarly, faculty may not want certain publications or grant data publicized because it is not representative of their current work or reveals too much about valuable intellectual property still being developed competitively. Some faculty do not want their photographs published. One post-doc objected to having her title and educational history listed.

Besides the policy problems these exceptions represent, there is a technical barrier currently in VIVO for maintaining exceptions so that they are not overwritten by the next ingest. Further, if the exception is temporary, how should it be flagged for correction later on? For the most part, we have received very little feedback that the Weill Cornell instance of VIVO is publishing sensitive or unwanted data. For this, we must credit the practice of omitting such data from source systems' output feeds and files. For example, Faculty Affairs does not provide data on historical members of the faculty, including the deceased or those who have left the institution.

VIVO's national Development Team has cautioned implementation sites against publishing any kind of sensitive information in a public-facing VIVO. Unlike in relational databases where given fields or tables can be hidden systematically, the semantic underpinnings of VIVO make this kind of security inherently difficult. Graphs of linked data are notoriously hard to partially restrict access. Given how effective VIVO is at defining relationships between disparate data and the infinite possibilities for how to query the data, we are now contemplating implementing a companion to our public VIVO, one that also contains sensitive or private data. Only selected, authorized, and authenticated users could access this VIVO. We could then more easily control programmatically what limited data is copied to the public VIVO.

5.4.3   PREFERENCE DATA

Possibly the most difficult policies surround how to handle data which varies by personal preference. For data with a high rate of variable preferences, programmatic solutions built into VIVO are essential. This would include preferred name and administrative titles. Preferences that are more difficult to represent include display sort order for a variety of fields, selective suppression of data like individual grants or publications, phone numbers, and email addresses.

5.4.4   WHAT CONSTITUTES A MINIMALLY POPULATED PROFILE

A year and a half into the grant and as our team was beginning to outline our outreach strategy, we agreed upon what constituted a minimally populated profile. We decided that the profile of every person with a current WCMC appointment and on the tenure, research, or academic-research tracks would be populated with at least the following types of data:

- education and training

- appointments/title

- heads of department

- agreement/grants (if the person has been named on an agreement/grant not involving vivisection)

- current and reasonably complete list of publications (if the person has authored a publication appearing in PubMed or Scopus)

We targeted these data because they were available. Also, these data are often, if not always, key elements of a CV or a biosketch. This is important because one of our goals for the project is to allow users to easily generate such documents using data from their VIVO profiles.

5.4.5 INCLUSION CRITERIA

In order to minimize user complaints, we aimed from the outset to be consistent about who would and would not be in VIVO. We started with directly employed faculty and sought to add more affiliated faculty, fellows, and graduate students over time with the ultimate goal of including staff and administration as well. In retrospect, the hand-wringing over who should go in our VIVO could have been minimized if we were more pragmatically system-centric, stating that our VIVO would contain:

- anyone named on a grant/agreement in our grants management system

- anyone who is a post-doctoral student, fellow, or person with a faculty appointment at Weill Cornell or any of the other CTSC institutions—*provided data was available from an official source system of record*

Over time, our thinking about representing researcher data in VIVO has evolved. We are now of the mindset that if we can ingest authoritative linked data relevant to our researchers, we will do so. The limiting factor is availability of authoritative and well-structured data. While this is a pragmatic policy, as self-editing features in VIVO improve, we will be challenged to return to our original list if source systems are missing people we would like to include.

5.4.6 REPRESENTING DATA ACROSS INSTITUTIONS

One of the strengths of VIVO as a Semantic Web application is the ability to link data across instances—hence, diminishing duplications and improving data quality. At Cornell University, there are now two VIVO instances, Cornell VIVO and our own, WCMC (Weill Cornell Medical College) VIVO. Researchers who are affiliated with WCMC have their names listed in both instances because they are also members of the university. VIVO allows us to avoid worrying about duplicating data or keeping two versions in sync by using linked data in Cornell VIVO to pull WCMC profiles from the Weill instance of VIVO. To achieve this capability, the WCMC namespace was added to the list of external namespaces in the Cornell VIVO instance. Once a name with a WCMC namespace is clicked by the user, the Cornell VIVO instance will open a WCMC VIVO page for the linked profile.

This capability is critical to our CTSC implementation (still in progress), which will be able to link to MKSCC and Hunter instances. While linked data vastly simplifies maintenance of authoritative data, the fact that one VIVO is representing data from another VIVO does present some nuanced challenges when trying to correct erroneous data. It may not be self-evident to the end user that changes must be made at system sources to which the VIVO CTSC team does not have access. Therefore, VIVO itself needs to provide some information about the data's provenance or other tools to help users report needed corrections.

Linked data also avoids some of the problems of cross-institutional differences in data definitions. For example, user identifiers will almost certainly be different. Happily, VIVO simplifies data sharing by employing a hierarchical and semantic data model in order to compare and combine data across institutions. However, even with linked data, the success of a query depends on knowing how any changes to local ontologies affects the results one gets.

Linked data also creates a mixed set of expectations regarding data update frequency. Each institution may have its own standard, and yet this may not be apparent to the user of a shared view. Similarly, harmonization of policies regarding what profile data should be displayed may be required.

5.4.7   FEDERATED AUTHENTICATION

Federated authentication is not required to use linked open data or to have a multi-institutional VIVO. However, some form of authentication is required to allow end users to log in to any given VIVO instance and edit their profile. This has been a challenge because although VIVO will operate with certain non-token authentication schemes such as LDAP,[5] their implementation exposes certain integrity risks for self-editing, and VIVO does not inherently support a less vulnerable mechanism.

We are currently exploring adding Shibboleth[6] capabilities to VIVO so that we can turn on user self-editing with more confidence. Though Shibboleth is generically designed for federation, initially this implementation will only allow WCMC users to edit. CTSC partners would be able to edit their profiles as they implement Shibboleth, and profiles hosted by other VIVO instances would need to authenticate by whatever method is used in their home institution's VIVO.

In addition to certain technical attractions, Shibboleth is the most promising federated authentication standard for VIVO because it is the basis for the InCommon[7] consortium, which already serves and is open to most of the research institutions for which VIVO is a desirable tool.

5.5   DATA INGEST

Working with a semantic application like VIVO required that we modify our approach typically employed with relational databases. For example, all the data in VIVO must be matched against the core ontology of classes and properties. Among the tools we used were VIVO Harvester and one we developed especially for this project, Google Refine+VIVO.[8]

---

[5]Lightweight Directory Access Protocol,
 http://en.wikipedia.org/wiki/Lightweight_Directory_Access_Protocol.
[6]http://en.wikipedia.org/wiki/Shibboleth_(Internet2)
[7]http://www.incommonfederation.org/
[8]http://bit.ly/GoogleRefineVIVO

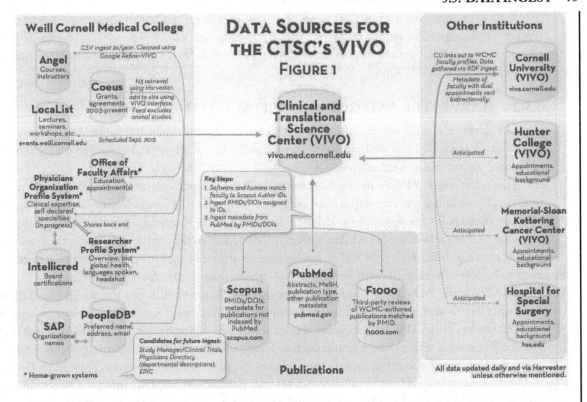

Figure 5.1: The CTSC's VIVO regularly ingests data from several different categories of sources: Weill Cornell Medical College (green), external sources providing publication metadata (pink), and other institutions (blue).

5.5.1   DATA MAPPING AND MODELING

Before acquired data could be ingested for use in VIVO, the team had to decide how to map the data to the VIVO ontology. While extensions to the VIVO ontology are allowed, there is a loss of functionality and increase in complexity if custom mappings are used rather than finding acceptable maps to the standard. Through an ongoing dialogue with other implementation sites, it has become much clearer how our metadata should be modeled in VIVO.

Getting clarification of how data is actually used from system owners and operators is critical to proper mapping. Some of our data sources have precise definitions for their source data while others do not, which leaves a small minority of types of data ambiguous. For example, there are several ways to semantically model a given clerkship course, particularly its date and time interval. Clerkship courses last for an inconsistent number of weeks or months, sometimes overlapping between one year and the next, and other times not.

As with many semantic models, definitions are not enough and highly interoperable standards require implementation guides that are currently lacking for VIVO. These should emerge with broader use and collaboration among users similar to the way IHE has developed implementation guides for HL7 and DICOM.[9]

5.5.2 HARVESTER AND GOOGLE REFINE+VIVO

VIVO Harvester is a library of tools designed to import data from authoritative data sources and ingest into VIVO. Because we have yet to turn on self-editing, all our data—except for local organizational names which were manually entered—are ingested through Harvester.

In addition to VIVO's native Harvester, WCMC has put extensive effort into developing methods for cleaning and importing data using Google Refine.[10] Google Refine is an open-source software package for manipulating grid-based datasets. Google Refine allows a dataset to be aligned to defined schema, to convert grid data into graph data, and to export the data into a triple format. Given these capabilities, Google Refine has become an attractive tool for working with datasets that need to be converted for use with Semantic Web applications, particularly "messy data."

In 2011, the members of our team,[11] spent several months building a Reconciliation API interface for VIVO, such that a given dataset containing dirty data may be seamlessly reconciled against a VIVO source. This VIVO reconciliation service acts as a Java HttpServlet that parses requests from Google Refine and returns query results back to Google Refine. With each request, it also returns a callback value in order to correctly map the results to the request that it receives. This allows Google Refine to use external systems to validate and enhance datasets. By implementing the Reconciliation API within VIVO, any VIVO instance could become a useful toolset for administrators working with institutional data.

This open source extension of Google Refine, known as Google Refine+VIVO, offers VIVO implementation teams a variety of valuable functions including:

- Ingest RDF from VIVO and use it as a starting point for data sets.

- Perform field-level validation of manually generated data or data extracts from other source systems against reliable data from VIVO.

- Enhance reconciled data with additional data columns from VIVO (email address, phone number, etc.).

- Prepare and clean data within Google Refine+VIVO. For example, this feature was useful for us in converting a spreadsheet of course data with inconsistent naming conventions into one that had such conventions.

- Align data to the VIVO schema.

[9]http://www.ihe.net/
[10]http://code.google.com/p/google-refine/
[11]http://www.vivoweb.org/files/VIVO%20Poster%20Final.pdf

- Export data to RDF to allow for easy ingest into a VIVO instance.

5.5.3   PUBLICATIONS METADATA

One of the liabilities of profiling systems that require manual entry (e.g., our legacy system) is that the most relevant publication tends to be the last one, the very publication least likely to be entered. Because publications are the means by which researchers communicate and are promoted and compensated—in essence, the lifeblood of a research institution—we made programmatically generated listings of publications an early priority, as did VIVO's national Development Team.

Determining which publications belong to which authors is a thorny problem. There are more than 22 million records in PubMed, and only a minority of co-authors has an assigned affiliation. The current PubMed Harvester only partially disambiguates all publications.

Within the terms of Cornell's existing license agreement with Elsevier, we have partnered with Scopus[12] to develop a semi-automated method for generating an up-to-date list of publications for Weill Cornell researchers. Our workflow, which also captures publications at our researchers' previous employments, is as follows:

1. Match researchers in VIVO to their respective Scopus Author Identifier(s).

   (a) Use Scopus's API to generate a spreadsheet-based reconciliation report of researchers with WCMC affiliation, any other affiliations they may have, and their respective identifiers.

   (b) Email to Scopus's Help Desk a spreadsheet indicating the approximately 1,100 cases where a person with a WCMC affiliation was clearly represented on two or more occasions. Scopus's team subsequently merged duplicated identifiers.

   (c) Use Google Refine+VIVO to programmatically match 2,450 authors by email address to one or more identifiers.

   (d) Use Scopus's web-based author search interface to manually match 1,396 people to anywhere between one and four Scopus IDs. This process took 27 person-hours. Another 846 people, many of whom are clinical faculty and therefore tended to publish less frequently, had no apparent Scopus ID. Finally, we characterized 149 people as uncertain, typically because their names were too ambiguous to make an initial judgment.

2. Include those Scopus ID(s) in individuals' VIVO profiles. Note that these identifiers are linked, so one can easily see which publications are associated with a given identifier.

3. Query Scopus by Scopus Author ID to retrieve a list of unique identifiers (always DOI,[13] very often PMID[14]) assigned to a given publication.

4. Using these identifiers, query PubMed for any available publication metadata, including Medical Subject Headings (MeSH), abstract, and indexing status.

---

[12]http://www.scopus.com/home.url
[13]Digital Object Identifier http://www.doi.org
[14]PubMed ID

5. If metadata is unavailable for a given publication (e.g., the publication is indexed by Scopus but not PubMed), ingest any available metadata from Scopus with the exception of abstracts.

6. Map publications to the relevant class in VIVO. We have configured the VIVO front end to display publications in order of recency within the following ordered categories: In Press, Research Articles, Reviews, Conference Papers, Editorials, Letters, Comments, and Other Academic Articles.

7. If the publication has a status of "In Process" or "In Data Review" within PubMed, we check back monthly to see if the publication has been fully indexed.

8. For publications that are the product of NIH-funded research, we also check back to see if the record has a PMCID (PubMed Central Identifier). The presence of a PMCID signals compliance with federal law mandating that the NIH-funded research be deposited in the open access database PubMed Central.[15]

Troubleshooting an incorrect list of publications for a given researcher requires that we first determine whether the profile has an incorrect or missing Scopus Author ID. More rarely, there is the case of the Scopus ID that is "half right," in which a given Scopus author profile commingles incorrect publications with correct publications (Scopus developers have purposefully erred on the side of splitting rather than lumping). In this case, researchers must provide feedback directly to our team. Such feedback is reflected in a subsequent ingest. Explaining and documenting these subtleties is an important element of our communications strategy as we are beginning to aggressively promote VIVO.

We expect that it is or will be possible to ingest publications metadata using Thomson Reuters' ResearcherID or the Open Researcher & Contributor ID (ORCID) in similar fashion.

5.5.4   TESTING

We have developed a series of easy-to-run checks to verify the following: Harvester is ingesting data, absence of duplicate data, different subtypes of a higher level class like organization are listed, data from each different system of record is loaded, the contact form works, SPARQL Query Builder functions, new user accounts can be added, etc. As we discuss later, we also use the SPARQL Query Builder itself to indicate the presence of certain red flags such as which departments fail to list a head. However, we have concluded that given the variable nature of most of the data, the only true method for ensuring the accuracy of a system is to have the system owners and operators regularly review the system, particularly before a new version or data update goes live. Such people are intimately familiar with how their system's data has changed and what to look for.

---

[15]http://www.ncbi.nlm.nih.gov/pmc/

5.6   SPARQL QUERY BUILDER

In version 1.2, VIVO's national Development Team incorporated a SPARQL Query Builder, which allows logged-in users to construct and execute complex queries of semantic data. We have found this WYSIWYG tool invaluable. Example queries we use simply for the maintenance of the system include:

- persons with a Weill Cornell appointment but without a Scopus identifier

- persons on the tenure, research, or academic-research tracks that are not associated with any publications or that lack educational history

- publications, grants, or activities within a certain time range

- publications associated with members who have an appointment in a certain department

The ability to execute such complex queries showcases the power of representing data in a semantic way. We are currently working on use cases for end users to take advantage of an easier-to-use future version of the SPARQL Query Builder for administrative queries and the like. Presently, we are focusing on reports for senior administrators hoping to track research productivity and areas of potential scientific collaboration with our new Technology Campus opening in NYC in 2017.

5.7   CUSTOM EXTENSIONS TO THE ONTOLOGY

Thus, far, the core VIVO ontology has successfully modeled most of our data (Table 5.1). Even so, there have been several occasions where we wanted to express our data in a more granular fashion. In all cases, our custom classes/properties were subclasses/ subproperties of existing classes/properties in VIVO core ontology. This practice is considerate of the need for third-party crawlers, particularly those belonging to any of the federated search networks such as vivosearch.org, which aggregate our data with other VIVO instances.

Until recently, we were expecting to create custom classes or properties for the following: clinical specialty, clinical expertise, board certifications, geographic focus, languages spoken, and more granular distinctions among publication types. However, we expect that many of these classes and properties will be fully represented in subsequent versions of VIVO core ontology.

5.8   SUCCESSES

Since VIVO has gone live, we have received a variety of positive feedback from members of the community. Many reported thinking the polished look and feel were an improvement over the legacy system. Our users tend to "get" the idea of linked data, expecting that if a person has an appointment in Anesthesiology, a click on that link will take one to a page that list other faculty with such an appointment.

Table 5.1: Custom ontology extensions employed in the CTSC VIVO

Field	Explanation or example	Namespace: class/property (class unless otherwise listed)	Child of
Center Wide Identifier	unique ID assigned to every employee at the Medical Center	wcmc:cwid (data property)	vivo:identifier
Office	e.g., Academic Affairs	wcmc:Office	foaf:Organization
Academic Program	e.g., MD/PhD Program	wcmc:Academic Program	core:Program
Special Faculty Position	e.g., Anne Parrish Titzell Professor of Neurology	wcmc:Special Faculty Position	core:FacultyMember
Clinical Trial Agreement	Industry-funded agreement to conduct a Phase I, II, or III trial	wcmc:Clinical Trial Agreement	core:Agreement
Sponsored Research Agreement	Industry-funded agreement to conduct basic science and observational research	wcmc:Sponsored Research Agreement	core:Agreement
Key Personnel Role	Besides the PI, people who contribute to the scientific development or execution of a project in a substantive way	wcmc:Key Personnel Role	core:Investigator Role
Key Personnel on	Inverse of key personnel role	wcmc:hasKeyPersonnelRole (object property)	core:hasInvestigatorRole
Personnel Role	People who are named on a project but neither a PI nor a key personnel	wcmc:Personnel Role	core:ResearcherRole
Personnel on	Inverse of personnel role	wcmc:hasAgreementPersonnelRole (object property)	core:hasResearcherRole

Perhaps the biggest coup has been the design of a workflow that will successfully ingest and display new publications independent of any need for user feedback or administrator action. We use these data to identify publications that are the result of NIH funding, thus greatly aiding in our efforts to ensure researchers are in compliance with NIH. We have performed an analysis of which institutions, including affiliates, our researchers collaborate with most. Also, we created a search strategy that lets interested administrators know when a publication appears in a journal of a given impact factor.

Another accomplishment has been that the College has publicized for the first time (most of) its agreements and grants data. Until we represented the data in VIVO, it was not at all clear which people or departments were working on which grants. Now, the user who calls the Library and asks for active grants being administered by the Department of Surgery can simply be shown how to find his answer in VIVO in a matter of seconds. Furthermore, the work required to define these relationships was comparatively minimal.

Going forward, we expect that allowing for export of VIVO data to a template like a CV or biosketch will elicit the greatest positive feedback from researchers. As we'll discuss later, we are just now plotting out a strategy for performing secondary analyses and outputs of the data in VIVO.

5.9 SELECT REMAINING ISSUES

5.9.1 CLINICAL SPECIALTY AND EXPERTISE

As previously discussed, WCMC uses IMO's interface terminology to allow physicians to easily assign user-friendly terms defining clinical specialties and areas of expertise to their legacy system profile. There are several issues that need to be addressed before we successfully model specialty and expertise within VIVO.

First, such terms may be employed in any of several contexts. We have conceptualized three distinct types of expertise in the clinical realm:

- **Personal sub-specialization**—This may be relatively obscure and highly specific. It is also what is most likely to differentiate Cornell's physicians from their peers. For example, Dr. X is a general surgeon who is world famous for burns. Dr. Y is not world famous but his interest and expertise in primary prevention techniques is far greater than that of his peer internists. Dr. Z spends most of his time reading echocardiograms, but he is best known for his expertise in Marfan's syndrome. This specialization is likely to correlate with publication history.

- **Core specialization**—This is the "bread and butter" expertise of any given physician. In the example above, Dr. Z's expertise in echocardiography would fall here. In some physician indices, this is called the physician's "typical patient." This specialization is likely to correlate with billing data.

- **Basic competency**—This is the general domain of the physician's specialty and sub-specialty. These are the hundreds of syndromes and procedures that any physician of a given specialty is qualified to treat, though they may not self-identify as experts in this area. Any internist can treat diabetes. Any general surgeon can perform an appendectomy. Relative to the patients they treat, they are clearly qualified experts. But each would, respectively, recognize an insulin researcher or inventor of a new laparoscopic technique as more specifically expert.

Compound terms present another real-world challenge to representing clinical expertise using standard terminology. Up to half of the terms selected by physicians—including "HIV Pulmonary Complications," "Pre-Conception Counseling," and "Psychotherapy and Cancer"—have no exact equivalent in even the most expansive controlled medical vocabulary (UMLS). The need for compound terms is also relevant to researchers with an interest in global health. For example, a typical global health expertise would be "tuberculosis in Haiti."

Thanks to work done at State University of New York at Stony Brook, version 1.4 of VIVO now features a UMLS lookup service allowing users to easily add UMLS terms to a VIVO profile. Our Implementation Team has yet to determine if we should:

- instruct users to use the UMLS service as is

- map terms from POPS to UMLS and ingest into VIVO

- duplicate POPS IMO-based term lookup functionality within VIVO

5.9.2 SELF-EDITING

At Weill Cornell, researchers and their staff must enter information into as many as 40 different informational and compliance systems. With VIVO, we wanted to minimize any additional burden and ensure that profiled researchers are asked to add data for which there is no authoritative or accessible data source.

Over the next several months, we intend to roll out VIVO's self-editing feature, using a single sign-on solution—InCommon. With self-editing, researchers and their proxies will be able to add data to select fields such as professional service, presentations, awards and prizes, and committee work to a profile.

For other fields such as appointments and education, we do not intend to allow researchers to edit. At this point, the plan is to streamline the process of sending feedback to the relevant system owners/operators.

5.9.3 TRANSITION TO OPERATIONS

We are presently in the process of transitioning the VIVO project from implementation to operations. As we plan this transition, a number of key issues must be satisfactorily addressed:

- **Personnel**—While the effort will vary, most of the same skills needed to set up VIVO remain important for the ongoing operations, including programming, system administration, ontology management, training, and system oversight.

- **Documentation and support**—For VIVO developers and others with administrative roles, we need to further document policies and procedures within the wiki space devoted to the project. The Service Desk will need to know how to troubleshoot common questions by end users. Further, we hope to provide answers to additional frequently asked questions within VIVO itself, some of which are video-based tutorials.

- **Outreach and training**—Particularly with the implementation of self-edit, we will need to regularly present to various groups of WCMC researchers about how to take full advantage of VIVO, including how they will be able to manage their profiles and generate reports.

- **Private VIVO**—We are currently considering creating a VIVO for administrators only as one that would include, along with the public-facing VIVO, more sensitive metadata such as animal studies, salaries, and size of agreements/grants.

ACKNOWLEDGMENTS

The content of this chapter represents the work of the entire WCMC VIVO team as well as the myriad contributions of the larger VIVO community that helped us implement. The authors would like to especially thank their colleagues at Cornell Ithaca and at the WCMC Clinical and Translational Science Center for their advice, support, and encouragement.

CHAPTER 6

Extending VIVO

**Chris Barnes, Stephen Williams, Vincent Sposato, Nicholas Skaggs,
and Narayan Raum,** *University of Florida*
Jon Corson-Rikert, Brian Caruso, and Jim Blake, *Cornell University*

Abstract

Since 2003, VIVO has evolved from a platform emulating Semantic Web concepts to become a set of tools built on Semantic Web standards and capable of leveraging ongoing advances in semantic technologies. In previous chapters, we have learned the motivations and context for VIVO, understood more about the principles underlying it, and observed how VIVO has been implemented at a range of institutions. This chapter will address the VIVO application with a special emphasis on how it can be modified and extended with the participation of an open community of developers. We will attempt to place our discussions in a larger context than VIVO itself, addressing challenges not limited to Semantic Web applications but extending to any project seeking to create networked production services in a time of rapid technology change.

This chapter will summarize VIVO's basic functions from a more technical vantage point and then discuss how VIVO has been extended by a larger development effort over the past three years and is now transitioning to a growing community development process.

Keywords

web development, user interface, Jena, Java, triple store, Tomcat, Apache, Solr, authentication

6.1 VIVO APPLICATION OVERVIEW: FUNCTIONS, COMPONENTS, AND TOOLS

6.1.1 KEY APPLICATION FUNCTIONS

VIVO provides three major functions within a single web application: an ontology editor for specifying the types of information to be modeled and the relationships among these types; a content editor for populating the ontology with data; and a searching and browsing interface for delivering VIVO content to end users. Close coupling of the content creation and editing environment with the customizable front-end display capability enables VIVO to support rapid prototyping of new types of content as needs expand or new data become available.

Local branding and content priorities as well as local extensions can make VIVO look and even behave quite differently in different installations. This flexibility is intentional, and it does not interfere with the ability to link data from multiple VIVO applications. As discussed in Chapter 2, most local ontology extensions "roll up" into core VIVO ontology classes and properties. The VIVO linked data index builder used for the VIVO integrated search demonstration (vivosearch.org[75]) makes few assumptions about the software delivering the data. One institution included in the demonstration uses a completely different software platform, Harvard Profiles, which shares the VIVO ontology as its internal data model;[76] other institutions, including the University of Iowa[77] and the University of Pittsburgh,[78] map data from internal systems to VIVO after export. Participation in any integrated VIVO search index requires only the ability to provide data expressed in the VIVO ontology in response to linked data requests that could be satisfied simply by placing RDF files in a web-accessible directory served by Apache.[79]

VIVO has always provided a straightforward presentation of information driven largely by the relationships inherent in the data and by a focus on search and discovery. Through the NIH-funded VIVO project, notable extensions to VIVO have provided a range of data ingest services and transformed VIVO's presentation and impact through the addition of the sophisticated visualization tools to be presented in Chapter 7.

6.1.2   VIVO OPEN-SOURCE COMPONENTS

VIVO has a layered construction built on a number of open-source tools and code libraries. As a Java web application, VIVO relies on the Tomcat servlet engine[80] for its primary operating environment. Most installations of VIVO place the Apache HTTP server[81] in front of Tomcat for handling incoming requests in order to take advantage of standard authentication modules that are available for Apache to support Kerberos[82] and Shibboleth,[83] the most common authentication tools used with VIVO.

Through version 1.4, VIVO relies on the open-source Jena Semantic Web Framework[84] to connect the web interface to a persistence layer, currently Jena's SDB triple store. VIVO may be configured to use several databases to host the SDB triple store; the open-source MySQL[85] database

---

[75]http://vivosearch.org
[76]http://profiles.catalyst.harvard.edu
[77]https://www.icts.uiowa.edu/confluence/display/ICTSit/Loki-Vivo+Alignment
[78]http://dl.acm.org/citation.cfm?doid=2147783.2147785
[79]http://www.w3.org/TR/swbp-vocab-pub
[80]http://tomcat.apache.org/
[81]http://httpd.apache.org/
[82]http://web.mit.edu/kerberos/
[83]http://shibboleth.internet2.edu/
[84]http://incubator.apache.org/jena/
[85]http://www.mysql.com/

is most commonly chosen, although the Amazon Relational Database Service[86] and Oracle[87] are also being used at several institutions. Version 1.5, released in July of 2012, will reduce VIVO's dependence on Jena and make it possible to connect to triple stores offering access via a SPARQL[88] query endpoint.

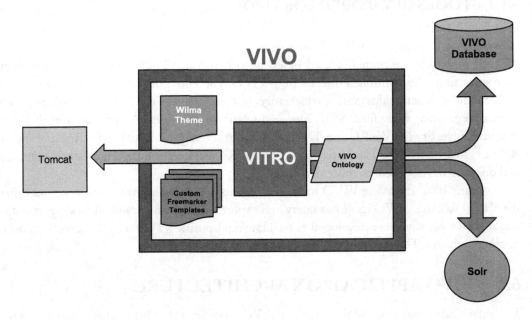

Figure 6.1: Overall components and flow of the VIVO application.

VIVO also uses a number of other open-source tools and services. The Apache Solr search server[89] provides an index used both for searching and for speeding up certain kinds of page displays and editing interaction within the application. VIVO has its own simple reasoner for data stored in its triple store but relies on the Pellet OWL 2 reasoner for Java[90] for in-memory reasoning on the ontology. The FreeMarker Java Template Engine[91] library has replaced much of the original reliance on Java Server Pages[92] within the application, and JQuery[93] and other JavaScript libraries provide key interaction support.

Beginning with version 1.4, VIVO has also adopted a layered distribution mode to permit adoption of its core functionality, called Vitro, in other application domains. Vitro is packaged with

---

[86]http://aws.amazon.com/rds/
[87]http://www.oracle.com/technetwork/database/options/semantic-tech/index.html
[88]http://www.w3.org/TR/rdf-sparql-query/
[89]http://lucene.apache.org/solr/
[90]http://clarkparsia.com/pellet/
[91]http://freemarker.sourceforge.net/
[92]http://www.oracle.com/technetwork/java/javaee/jsp/index.html
[93]http://jquery.com/

the VIVO application or may be downloaded and used independently of the VIVO ontology for teaching Semantic Web concepts, developing new ontologies and populating them with data for prototyping or evaluation, or providing the platform for new Semantic Web applications.

6.1.3   TOOLS DEVELOPED FOR VIVO

Data ingest has been a major focus of the national VIVO project and of the development team at the University of Florida. The VIVO Harvester provides a comprehensive framework for data ingest through tools that are primarily used independently of the VIVO production system but which have some functions controlled directly from VIVO. The Harvester offers an extensible framework with modular functionality and a wide range of templates for different data sources, including comma-separated- value files, XML files, and relational databases. Additional contributions from Indiana University and Weill Cornell Medical College leverage tools such as D2RMap[94] and Google Refine[95] to augment the Harvester tools. Section 6.4 will describe the Harvester design, features, and ongoing development in more detail.

Many institutions use VIVO to collect, organize, and disseminate data in venues other than the VIVO website. VIVO itself has query and export mechanisms designed to support data reuse, and several tools have been developed to facilitate repurposing VIVO data for reporting and for use in other websites. These tools are summarized in Section 6.4.

6.2   VIVO APPLICATION ARCHITECTURE

The extensibility and much of the power of VIVO stem from the ability of the application to adapt to changes in the ontology while maintaining direct interoperability with other Semantic Web data and tools. Extending the VIVO data model to represent new types of data and new relationships is quite simple, although it merits considerable thought, as discussed in Chapter 2.

However, this very flexibility on the data modeling side of the application can make extending the application to provide new controls or presentation features more complex. Modifying or extending VIVO can be daunting without orientation to the principal parts and functions of VIVO and how these components interoperate in tasks such as delivering web pages, responding to search queries, and providing edit controls. In many cases, the application makes use of secondary RDF models to manage key functions, including user accounts, permissions, menu pages, and editing and display controls.

6.2.1   VITRO

It will be easier to understand how VIVO functions by describing the anatomy and basic functions of Vitro itself and then showing how VIVO extends and customizes this core code. A brand new installation of Vitro is typically configured during the installation process to connect to a newly

---

[94]http://www4.wiwiss.fu-berlin.de/bizer/d2rmap/d2rmap.htm
[95]http://code.google.com/p/google-refine/

created and completely empty database and to have an arbitrary default namespace (e.g., `http://vitro.example.edu`) for data yet to be created and stored in the application. Behind the scenes, Vitro has created an empty Solr search index and populated a private RDF model of users with the default root administrative account; the lone default Vitro ontology has only two utility classes required to support image upload. There's no data to search or browse, and there is only a Home menu tab for navigation.

The normal first step in making a new application is to create the beginnings of an ontology by specifying its web-addressable namespace (e.g., `http://vitro.example.edu/ontology/`) and a shorthand prefix, and then adding a few classes and properties. Vitro expects that at least one class group will be set up to support grouped listings in search results and on the Index page, which is a list of all populated classes in the application. Once these first steps have been taken, the application generates a default editing interface for adding new content, and it displays that content using a default page template, showing any connections between entries as internal links.

As shown in Figure 6.2, the Vitro (and VIVO) startup process sets up an in-memory Java servlet context for the application where a number of configuration details are established for use in servicing incoming HTTP requests. Each new request that the application receives is routed through servlet filters to a controller determined by the URL of the request; the controller processes the request using information from the request headers, the servlet context, and the current user's session to prepare a response. Depending on the nature of the request, the response may include accessing the application's triple store through a common data access object (DAO) layer, processing the data retrieved using methods in Java, or processing the results with one or more Freemarker page templates. Freemarker templates help Vitro and VIVO separate data processing or business logic from display logic; templates may include other templates as in any typical web-application framework where page headers, footers, and other composition elements are reused on multiple pages.

Straight out of the box, the Vitro software does not differentiate display of one type of data from another since every page in the browsing interface is generated using the same page template. Vitro offers a set of basic display controls that affect all instances of a given class or property uniformly; the ontology used, whether created in Vitro or imported, and the degree to which different classes and properties have been populated may cause significant variation in appearance among pages. Unless a user is logged in, only the properties that have been populated for a given individual are shown; properties may optionally be assembled into property groups and ordered within groups for optimal display.

Vitro's search functionality is managed via controllers that connect with the Apache Solr search engine packaged with Vitro and the more fully featured VIVO application; results are clustered and sorted via membership of classes in class groups.

Authentication and authorization comprise the most significant additional core functions of Vitro. The initial root user may create other user accounts and assign role levels to those accounts using a set of pre-configured roles in Vitro, ranging from a self-editor with the fewest privileges up

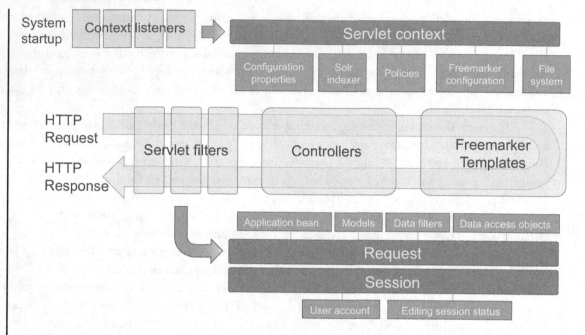

Figure 6.2: Overview of the Vitro application architecture from the perspective of a start up and HTTP Request.

through editor, curator, and administrator. Users who do not have accounts have no editing privileges; the privileges of self-editors and their designated proxies are governed by a policy framework in the Vitro application that, by default, implements only the basic role levels mentioned above.

While Vitro offers a convenient combination of features in a single package, the application is so generic in appearance that most adopters will want to do some customization, at least of the application theming and branding. Vitro includes a simple visual theme that may be extended by copying the entire theme directory, making modifications in the copy, and then specifying the modified copy as the preferred theme on the Vitro site administration page.

6.2.2   VIVO: THE FIRST EXTENSION

The VIVO application extends Vitro by providing a fully developed ontology and a large number of code modifications and additions, many of which leverage the VIVO ontology in delivering additional functionality. Modifications developed to date indicate paths to follow for further customization or extension of VIVO to meet local requirements.

Some of the changes in VIVO are major—for example, the entire range of visualizations encompassed in VIVO and described in Chapter 7 are dependent on the VIVO ontology, so they require the full VIVO application. Only VIVO has integrated Harvester functions and QR code display, both of which also depend on the VIVO ontology. Other additions to VIVO have accrued over many years of development at Cornell and, more recently, via contributions from the wider development community. Examples include additional menu pages, custom templates for people versus other types of entities, custom data entry and editing forms, and a more elaborate visual theme.

6.3   TYPICAL VIVO MODIFICATIONS

6.3.1   THEME CHANGES

A theme in VIVO is a collection of files that together define the behavior and presentation of the VIVO application and the content in a VIVO installation. A theme contains three major types of files:

- Freemarker template pages that are combined with data to produce output HTML to display on a VIVO page;

- CSS style sheets that govern the appearance and placement; and

- JavaScript that supports context-specific interactivity.

These files are organized into a directory structure that includes a mandatory *templates* directory.

Customizing a theme starts with copying the entire default theme (wilma) to the themes directory in the local application source directory, and then specifying a different theme name. Switching VIVO to use the copy as its active theme causes the application to look there to find an override for an existing template or any new templates added as part of customizing the appearance of the application.

The Site Administrator Guide on the VIVO SourceForge wiki includes extensive documentation on modifying the default theme.[96]

6.3.2   ONTOLOGY EXTENSIONS

As described in Chapter 2, the VIVO ontology provides semantic definitions to the data classes and properties, allowing intelligent inferencing and the ability to create connections between new data. Ontology extensions in many cases do not require any changes to the VIVO software or configuration, and can be effected using the native Vitro ontology editor. For example, most institutions have an internally defined, unique institutional identifier for each employee, and a new data property will be added specifying the name of the identifier as its label. If the property has the class foaf:Person assigned as its domain, and if it is set to be editable, the property will appear for editing with every foaf:Person entered into VIVO. The application allows configuring the level of user for which the property is displayed, the group of properties to associate it with, and the display order of the property within the group.

[96]http://sourceforge.net/apps/mediawiki/vivo/index.php?title=Site_Administrator_Guide#Developing_VIVO_Themes

6.3.3   CUSTOM LIST VIEWS

In Vitro, individuals introduced as instances of specific classes are labeled with a name and an additional tagline of the public label for the class; for example, if the ontology defines a class called "Health Care Facility," the entry for a hospital would be displayed as, "Marshall Community Hospital | Health Care Facility."

The public label on a data or object property also governs the default property display in Vitro and VIVO; if in the example above a data property called "number of beds" has been defined and populated for the health care facilities that have been entered, each would show a "number of beds" property and the appropriate value. If an object property called "located in" has been defined to connect the facility to a member of another class called "Municipality," and if appropriate data has been entered, Vitro and VIVO will by default show "located in" and the label of the related municipality. In most cases this default behavior is adequate, but sometimes a customized display of information on types and relationships can dramatically improve the usability of the application.

A custom list view is a small, localized extension that uses a SPARQL query to retrieve additional information to display and/or executes some logic to display different information under different conditions. Writing a custom list view is not difficult, but it does require knowledge of the ontology, the SPARQL query language, and the syntax used in Freemarker templates—or at least a willingness to modify an existing example to suit local needs. Perhaps the most involved custom list view shipped with VIVO is the publication list view that assembles and displays a full publication citation for each publication listed on an author's page in VIVO. This custom list view executes a set of queries to return the names of all related authors, the journal associated with an academic article, the book within which a book chapter was published, and other bibliographic information, together with logic to determine how to display a book chapter differently from an academic article and what to do when one or more parts of the citation are missing.

Another very useful example has recently been added to a multi-institutional VIVO implementation where not only a person's department needs to be displayed for each position he or she holds, but also the department's institutional home. Without the additional indicator of home institution, the department reference alone is not unique. In this case, the query implemented for the custom list view needs to be able to search recursively for each successive parent organization of the department until a top-level institution is reached.

6.3.4   CUSTOM TEMPLATES

An out-of-the-box Vitro implementation uses the same page template for every type of data entered—great for consistency, but not ideal for applications designed to optimize presentation based on context and content. VIVO introduces an additional page template for people that significantly extends the default template by highlighting research areas, appointments, contact information, and related websites at the top of the page and by grouping other properties below. In large part, VIVO page templates have been brought into compliance with model-view-controller

(MVC) design principles[97] with releases 1.2 through 1.4 and the switch from Java Server Pages[98] to Freemarker[99] templates, but more nuanced presentation typically also involves more extensive use of JavaScript libraries and Ajax[100] or HTML5[101] development techniques to deliver responsive interactive behavior. Many page templates also include other template fragments in the interest of code reuse and reliability, with the tradeoff being that chains of imports can be complicated.

6.3.5   MENUS AND PAGES

In addition to data display pages, other common page types include menu pages, index pages, and search results. With VIVO 1.4, menu pages became much more configurable; VIVO 1.5 supports adding arbitrary static HTML pages and pages that include the results of one or more SPARQL queries that may optionally include user-specified parameters. The former facilitates providing arbitrary additional background information about a VIVO implementation or its host institution, while the latter puts a very powerful capability in the hands of site administrators willing to write SPARQL queries and format the results. Adding menu pages populated by arbitrary SPARQL queries enables sites to create and share reports highlighting recent content such as new grants or publications or networking features such as collaborations among different colleges within a research university. The implementation community has already begun sharing useful SPARQL queries for purposes ranging from estimating the extent of profile population to checks for orphaned data that has lost any connection to other data in VIVO.

6.3.6   LOGGING ACTIVITY IN VIVO

A number of VIVO sites have requested the ability to track additions, modifications, and removal of content as a way to understand how VIVO is used, how many users are editing their own profiles, which types of data are most frequently or infrequently modified, and the amount of data brought in by automated ingest processes. The University of Florida has implemented listeners within the application that write out daily logs of VIVO activity, including the name of the user or automated process making the change, the date and time of the change, the nature of the change, and a copy of the actual data added or removed.

These logs add considerable management oversight to VIVO while not materially affecting the responsiveness of the system or adding unnecessarily to the VIVO database through more complex means such as reification[102] that involve layering additional statements about VIVO triples in the triple store. The files are stored in a web-accessible directory using a simple comma-separated-values (CSV) format for ease of analysis using standard spreadsheets and other common tools. Modifications of VIVO to support logging are scheduled for the VIVO 1.6 release.

---

[97]http://st-www.cs.illinois.edu/users/smarch/st-docs/mvc.html
[98]http://www.oracle.com/technetwork/java/javaee/jsp/index.html
[99]http://freemarker.sourceforge.net/
[100]http://www.adaptivepath.com/ideas/ajax-new-approach-web-applications
[101]http://en.wikipedia.org/wiki/HTML5
[102]http://www.w3.org/TR/rdf-mt/#Reif

6.3.7 ALTERNATIVE AUTHENTICATION PROTOCOLS

While VIVO supports a database of users and can manage all aspects of their accounts including passwords, it is advantageous at most institutions to connect VIVO to the institutional single sign-on authentication. Users don't have to create or maintain another password, and VIVO site managers don't have the headache of providing support associated with passwords.

Most institutions we have encountered not only maintain unique identifiers and a single sign-on infrastructure for their employees (and students, in the case of academic institutions), but they also provide any necessary tools for using that infrastructure in other applications within the organization.

In Cornell's case, this infrastructure includes a locally maintained authentication module that can be deployed with Apache web servers[103] used by Cornell for its single sign-on system.

The University of Florida uses a similar approach, but with the Internet2 Shibboleth single sign-on identity management system.[104] The Apache authentication module is provided as part of the Shibboleth distribution, and it is configured to work with the University of Florida's central Shibboleth server.

With either identity management system, a special URL within VIVO is designated to the Apache module as a secure folder; Apache handles routing the user's authentication request to the institutional authentication system and recognizes additional attributes included in the HTTP headers. If authentication has been successful, these additional header attributes include the unique institutional identifier of the user. If the VIVO installation has been configured with the name of the property used for the institutional identifier, VIVO will match the value of this property with the attributes included in the request header and associate the user's session with his or her own data and display page in VIVO.

Once VIVO knows who is editing, the appropriate policies can be invoked based on a person's role level in VIVO to control what can be edited. A typical "self-editor" with no special privileges in VIVO can only edit information associated directly with him or herself and information related via designated properties such as authorship or investigatorship on a grant.

The American Psychological Association is working with VIVO to allow authors at VIVO institutions to log in to vivo.apa.org using their home university credentials and claim their publications. Using the InCommon Shibboleth tools,[105] the APA and universities will then be able to exchange trust assertions and provide visible confirmation of authorship in VIVO at either end. This initiative is part of a larger APA initiative to develop a scalable, trust-based scholarly publishing infrastructure known as the Publish Trust Framework.[106]

---

[103]https://identity.cit.cornell.edu/authn/
[104]http://shibboleth.internet2.edu/support.html
[105]http://www.incommonfederation.org/
[106]http://www.publishtrust.org

6.3.8   EXTENSIONS ACHIEVED THROUGH VIVO MINI-GRANTS

The VIVO project sponsored a mini-grant competition in 2011 to encourage VIVO adoption and promote development related to VIVO.[107] Funded projects include features that have been incorporated into VIVO and development that is continuing to evolve independently of VIVO.

Stony Brook University developed a web service for the National Library of Medicine's Unified Medical Language System[108] as a VIVO mini-grant and has continued to host the service for use by VIVO and other applications; as described in Chapter 2, a terminology search and review interface was added to VIVO 1.4 to support this service and others offering a similar lookup capability for different disciplines. By referencing established vocabularies with stable URIs rather than creating new VIVO-specific URIs or simply storing string representations of terms, data from one VIVO will be that much easier to align with data from another VIVO or from any source of linked open data on the web also tagged with the same term.

The University of Pittsburgh used VIVO mini-grant funding to spin off Digital Vita Documents (DV-Docs) functionality from the Digital Vita research networking system[109] and develop a mapping from the VIVO ontology to the Digital Vita schema. As of version 1.3, VIVO includes for its part a rich export format that returns all the data for a person's curriculum vitae in response to a single request; DV-Docs transforms the RDF to XML and uses the Apache FOP library to output a CV in rich text[110] or PDF format. DV-Docs also supports NIH biosketch formats.

A mini-grant to Open Researcher and Contributor ID (ORCID)[111] explored adding a Cross-Ref lookup for publications within VIVO,[112] a standalone Ruby-on-Rails[113] application[114] sharing VIVO profile data using the OAuth[115] protocol for secure API-level authorization between applications. A project with the Indiana Clinical and Translational Sciences Institute[116] developed a SPARQL query builder to browse content in VIVO and pull faculty profile information across to the CTSI HUBzero portal.[117] The Duke VIVO Widgets project[118] enables repurposing VIVO data outside of the application, including an application that allows users to search content in VIVO from their OpenSocial container, such as an iGoogle page.[119]

Independently of VIVO, Eric Meeks and colleagues at the University of California–San Francisco and other institutions developed OpenSocial gadgets designed to work with RDF expressed using the VIVO ontology.[120] OpenSocial is an open standard defining a web-based container envi-

[107] http://sourceforge.net/apps/mediawiki/vivo/index.php?title=VIVO_Mini-Grant_Projects
[108] http://www.nlm.nih.gov/research/umls/
[109] https://digitalvita.pitt.edu
[110] http://en.wikipedia.org/wiki/Rich_Text_Format
[111] http://about.orcid.org/
[112] https://github.com/gthorisson/vivo-orcidextensions/tree/master/productMods
[113] http://rubyonrails.org
[114] https://github.com/gthorisson/vivo-orcidextensions/tree/master/demoApp
[115] http://oauth.net/
[116] http://www.indianactsi.org/
[117] http://uits.iu.edu/page/axcg
[118] http://sourceforge.net/apps/mediawiki/vivo/index.php?title=VIVO_Widgets
[119] http://www.google.com/ig
[120] http://www.slideshare.net/ericmeeks/vivo-2011-profiles-open-social-rdf

ronment for hosting third-party components in a web application and a set of common application programming interfaces for developing these components[121] by leveraging the Google Gadgets framework.[122] The first gadgets have been implemented in Harvard Profiles, a researcher networking system developed by Griffin Weber at Harvard and modified in 2011 to use the VIVO ontology as its native data format.[123]

VIVO has been extended to serve as an OpenSocial container with version 1.5, allowing VIVO to share gadgets with Harvard Profiles and potentially other software systems supporting the VIVO ontology in the future. Gadgets can typically be added to existing pages or displayed as full pages in their own right, depending on the amount of data and screen real estate required. The vision for VIVO-compatible OpenSocial gadgets is to allow third-party developers to extend VIVO or Profiles to meet requirements in their local context—such as providing support for medical students to find mentors among the faculty—while also making that functionality available elsewhere with minimal modification. The relative ease of development and deployment of gadgets is seen as a lightweight way to extend functionality across a distributed community.

6.3.9   MODULARITY

There are several motivations for increasing the modularity of the VIVO application. These include being able to replace one part of the application with a preferred alternative implementation, being able to run a VIVO without unwanted features, providing a clear path for adding new functionality, and segregating code to avoid unanticipated side effects from changes in existing code.

VIVO already has a high degree of data interoperability through its adherence to Semantic Web standards for its ontology and data storage layers. VIVO's employment of the Jena Semantic Web framework[124] allows implementation sites to use any database supported by Jena, including MySQL,[125] PostgreSQL,[126] and Oracle.[127] VIVO Release 1.5 allows the use of triple stores that support reading and writing data through a SPARQL endpoint, although testing will be required to verify full compatibility and acceptable performance.

The VIVO Harvester uses quite a different application architecture and may be installed independently of VIVO; the VIVO visualization suites have been designed to separate VIVO-dependent components from the visualizations themselves in order to facilitate reuse in other applications.

There are many aspects of modularity and extensibility that VIVO does not yet address, however. Extensions should be simple to add, remove, and configure in the application. There are many overlapping dependencies in the VIVO application, and it would be helpful to separate them into those required at build time versus those that take effect only at runtime. In reviewing options for improving configuration and deployment as well as modularity and extension, the VIVO development

---

[121]http://docs.opensocial.org/display/OS/Home
[122]https://developers.google.com/gadgets/
[123]http://profiles.catalyst.harvard.edu
[124]http://incubator.apache.org/jena/
[125]http://www.mysql.com/
[126]http://www.postgresql.org/
[127]http://www.oracle.com/technetwork/database/options/semantic-tech/index.html

team has elected to use technologies developed by the OSGi Alliance[128] as a Java framework that is well understood and offers several alternative implementations. Preliminary investigations conducted during the version 1.5 development cycle have been positive, and tangible changes involving an embedded OSGi framework are expected in version 1.6.

6.4   TOOLS FOR DATA

The VIVO development team has been keenly aware of the necessity to provide the full range of tools needed to add, edit, and curate all the data and relationships necessary to make a VIVO function and thrive. These include, but are not limited to, tools for cleaning data prior to ingest, data ingest tools such as the VIVO Harvester, and tools that work with data already in VIVO to discover malformed or orphaned data and/or remove duplicates.

It's also critical that applications storing and producing data that has intrinsic value to an institution over the long term include robust functions for exporting that information in a format that can be migrated forward and reused in other applications. VIVO itself allows export of both its data and the ontology, and the VIVO community has produced a number of tools to support data reuse in other formats as well.

Entering data by hand can take a long time, and most institutions already maintain a number of data sources recording information on employment, grants, courses, events, and news. One role VIVO often plays is the integration of multiple sources of data, using either its internal RDF-focused data ingest tools (described in Chapter 3) or the VIVO Harvester. The Harvester largely operates independently of VIVO itself and is designed to support repeatable and schedulable ingest processes associated with a production VIVO instance.

6.4.1   VIVO HARVESTER

The Harvester is a Java Library of small tools designed to work sequentially to fetch data from a remote source, translate it into RDF expressed with the VIVO ontology, match it against VIVO data, and add the resulting output into VIVO's data store. The Harvester uses the same Jena toolkit as VIVO itself; since Jena is written in Java, that platform was chosen for the Harvester as well.

The data-fetching process pulls data from PubMed, CSV files, relational databases, or OAI harvest and uses XML as an intermediate format prior to transformation through XSLT into the XML/RDF compatible with the VIVO ontology. A harvesting script must be configured for each new source in converting to XML; XSLT transformations have been developed for each of the major types of data (human resources, publications, grants, and courses) but may also need to be changed due to differences in the availability of information at institutions adopting VIVO. Several institutions have shared new configurations for fetch processes and offered improvements to the configurability and the data harvesting code itself.

[128]http://www.osgi.org/About/HomePage

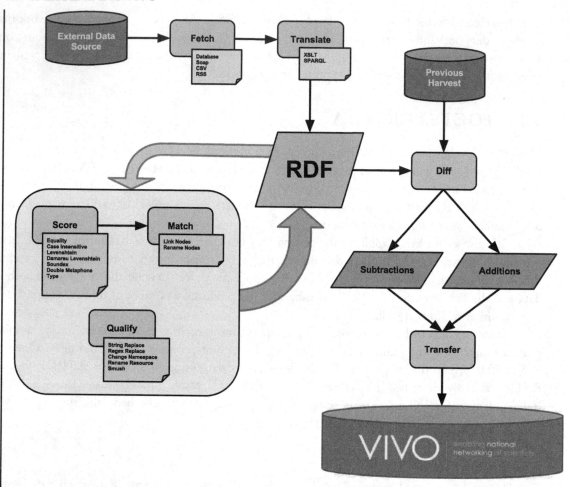

Figure 6.3: The basic workflow of a data ingest using the classes of the VIVO Harvester.

The matching process is more complex. The process of adding new journal articles to VIVO requires: matching the title and other metadata of the article itself against existing articles in VIVO; matching the journal by ISSN number, name, and/or abbreviation; and finally matching each listed author against people already in VIVO. Where a definitive identifier exists in both target and source, matching is simple, but some journals have more than one ISSN number for electronic and print editions, and not all article data includes the DOI information most frequently used for online linking. Authors present the biggest challenge for matching since often very little is known about them, especially before affiliation or email addresses were captured for authors other than the designated primary contact (normally the first author).

For these reasons, the scoring module of the Harvester continues to evolve as techniques such as string comparison using the Levenshtein distance algorithm[129] or Soundex indexing system[130] are implemented to address needs identified by sites using the Harvester, often employing a tiered evaluation to reduce the number of comparisons and a system of weights to determine the final score for each match. As with any matching effort, the threshold for acceptance can be somewhat arbitrary and involves trading the risk of false positives against extra hand-curation and/or the perception of low output data quality for seemingly obvious matches. The Cornell VIVO team has been developing a web-based post-ingest curation tool to identify close matches and provide an efficient interface for non-programmers to accept or reject proposed matches; this approach has been required in part because publication data has been derived from multiple sources, including a separate faculty reporting tool that relies heavily on manual data entry.

The final Harvester step transfers the transformed and matched data into the VIVO database using the same Jena Java libraries used by VIVO itself. A rewrite of the low-level VIVO data source interfaces in version 1.5 will allow the Harvester in future releases to call VIVO services for data transfer, which will have the added advantage that reasoning and search index updates will be triggered automatically.

6.4.2 DATA SHARING AND REUSE

The VIVO community has developed multiple tools for sharing data from VIVO—both tools that use VIVO's linked open data support (and extensions of that), and tools that leverage a SPARQL endpoint serving VIVO data. While VIVO has an internal SPARQL query capability for administrative and reporting use, and could be configured to expose an endpoint to the web, exposing the VIVO application to arbitrary queries from unknown parties is not recommended any more than it would be to expose an open SQL interface to a relational database. The Joseki[131] SPARQL server and its successor, Fuseki,[132] have been developed to serve RDF data efficiently over HTTP, and may be configured to query the same triple store used by a VIVO instance (currently SDB[133]) or any triple store that VIVO data can be replicated to, further isolating performance or security issues from VIVO itself.

Since VIVO itself responds to linked data requests, no additional infrastructure is required to support applications designed to access data via linked data requests. The VIVO Linked Data Indexer[134] is a utility application written to populate the Apache Solr search index used by the demonstration multi-site VIVO search, and uses linked data requests as a common-denominator approach that would work with any linked-data-capable application. The consuming application must determine how to process the data returned, including how to request additional information

---

[129]http://www.merriampark.com/ld.htm
[130]http://www.archives.gov/research/census/soundex.html
[131]http://www.joseki.org/
[132]http://incubator.apache.org/jena/documentation/serving_data/index.html
[133]http://incubator.apache.org/jena/documentation/sdb/index.html
[134]http://sourceforge.net/p/vivo/tools/code/1324/tree/LinkedDataIndexer/

if a first request does not provide all the information needed. Given the atomic structure of RDF, assembling all the information about an entity represented initially as a single URI could potentially involve hundreds of separate requests, and it's very difficult when harvesting RDF to know when to stop, given the web of relationships that may be represented in an RDF graph structure.

VIVO helps applications manage this by providing an enhanced form of linked data response that includes more labels than required and traverses common bridging relationships typified by the context node structure of the VIVO ontology, as described in Chapter 2. It will normally be faster and easier for another application to query a VIVO data store versus making multiple linked data requests and stitching together the results. As mentioned previously in Section 6.3.8, VIVO also has a rich export capability designed to support providing all the information expected for a curriculum vitae of a person via single request; this function was designed to be readily extensible through placement of additional named SPARQL queries in a directory visible to the application at runtime.

Three existing APIs for SPARQL endpoints are worth mentioning. The first is the Linked Data Import module[135] developed for the Drupal open-source content management system[136] by Miles Worthington. This module will work either via linked data requests or a SPARQL endpoint and uses the Arc2 RDF framework[137] to retrieve and parse RDF, while itself providing mapping to standard Drupal data structures.

The Semantic Services[138] developed by John Fereira are used at Cornell to syndicate VIVO content in JSON or XML formats to campus websites not yet capable of consuming RDF directly. A new open-source Linked Data API[139] offers a more general solution and has stimulated the development of at least two reference implementations—Elda[140] in Java and Puelia-php.[141] These APIs are all designed to provide interfaces and data familiar for web developers with no training or background in RDF, and can allow wide use of VIVO content outside of VIVO itself.

6.4.3 VIVO MULTI-INSTITUTIONAL SEARCH

The potential for services leveraging VIVO and its ontology across multiple institutions is best illustrated by the prototype search website developed for the 2011 VIVO Conference. Vivosearch.org demonstrates a unified search interface across the seven VIVO partner sites and Harvard Profiles, using data harvested with the Linked Data Indexer described above and processed into a single Apache Solr search index exposed via a Drupal website. Figure 6.4 shows the first page of results from a search across all eight sites for the term, "recombinant."

The vivosearch.org site is not a federated solution sending information and retrieving data at query time from separate applications; instead, data are harvested and compiled into a common

---

[135] http://drupal.org/sandbox/milesw/1085078
[136] http://drupal.org/
[137] https://github.com/semsol/arc2/wiki
[138] http://semanticservice.svn.sourceforge.net/
[139] http://code.google.com/p/linked-data-api/
[140] http://code.google.com/p/elda/
[141] http://code.google.com/p/puelia-php/

Figure 6.4: Search results on http://vivosearch.org showing faceting by institution and result type.

index for rapid response times and the ability to perform relevance ranking as well as faceting of results.

The potential impact of a multi-institutional search and likely scenarios for implementation and sustainability will be discussed in Chapter 8.

6.5   THE VIVO OPEN-SOURCE COMMUNITY

Source code was available from the start of the NIH-funded VIVO project via a BSD open-source license.[142] Following the first official release under the project in February of 2010, the University of Florida development team took the initiative to establish a VIVO project space on SourceForge and leverage the project space, Subversion source control system, mailing lists, hosted wiki, and issue tracker. The UF team has also supported an Internet Relay Chat (IRC)[143] channel for real-time interaction among implementers and developers.

This transition from development hosted on local servers to a much more visible venue was instrumental in demonstrating the VIVO project's intent to expand into a full open-source community at the first VIVO Conference in August 2010 in New York City. The Cornell and Indiana University development teams followed suit and merged Subversion repositories into SourceForge after the mid-year development meeting in February 2011.

Neither the development community nor the tools available for shared development stay static, and the VIVO community continues to explore other options for shared development, including Github.[144] The Harvester development team has recently migrated development to Github and begun using pull requests[145] as the preferred mode for distributed developers to submit code back to the project.

A successful open-source community must provide value to participating developers and to organizations supporting participation by their staff; the investment of time must be warranted by faster, easier, and better outcomes. Much of the benefit stems as much from communication as from direct technical help with questions like: Am I looking in the right place? Is my approach in line with others? How should I interpret this message? As the confidence of community members grows, are interacting directly with each other, without mediation by VIVO project staff. They share frustrations as well as excitement, encourage each other, and they provide a backup when someone has a deadline or just feels lost.

The strategic and organizational aspects of growing the VIVO community, including the community around the VIVO ontology itself, will be discussed further in Chapter 8.

ACKNOWLEDGMENTS

We'd like to thank all of the developers who have worked on VIVO in the past and all who will work on VIVO in the future. These contributions enable VIVO to be a tool for researchers around the world.

For those with the longest history with VIVO, the involvement of new people and the ideas they bring—including pushing for fully open development and planning—has been the most satisfying.

---

[142]http://www.opensource.org/licenses/BSD-3-Clause
[143]http://en.wikipedia.org/wiki/Internet_Relay_Chat
[144]https://github.com/
[145]http://help.github.com/send-pull-requests/

REFERENCES

[1] Devare, Medha, Jon Corson-Rikert, Brian Caruso, Brian Lowe, Kathy Chiang, and Janet McCue. 2007. "VIVO: Connecting People, Creating a Virtual Life Sciences Community." *D-Lib Magazine* 13 (7/8). DOI: 10.1045/july2007-devare

[2] Jeffery, Keith. 2010. "The CERIF Model As the Core of a Research Organization." *Data Science Journal* 9:CRIS7–13. J-STAGE. DOI: 10.2481/dsj.CRIS2

[3] Krafft, Dean, Nicholas A. Cappadona, Brian Caruso, Jon Corson-Rikert, Medha Devare, Brian J. Lowe, and VIVO Collaboration. 2010. "VIVO: Enabling National Networking of Scientists." In *Proceedings of the WebSci10: Extending the Frontiers of Society On-Line, Raleigh, NC, April 26–27*.

[4] Schleyer, Titus, Brian S. Butler, Mei Song, and Heiko Spallek. 2012. "Conceptualizing and Advancing Research Networking Systems." *ACM Transactions on Computer-Human Interaction* 19 (1): 2. DOI: 10.1145/2147783.2147785

[5] Weber, Griffin M., William Barnett, Mike Conlon, David Eichmann, Warren Kibbe, Holly Falk-Krzesinski, Michael Halaas, Layne Johnson, Eric Meeks, Donald Mitchell, Titus Schleyer, Sarah Stallings, Michael Warden, Maninder Kahlon, and Members of the Direct2Experts Collaboration. 2011. "Direct2Experts: A Pilot National Network to Demonstrate Interoperability Among Research-networking Platforms." *Journal of American Medical Informatics Association* 18:i157-i160. DOI: 10.1136/amiajnl-2011-000200

CHAPTER 7

Analyzing and Visualizing VIVO Data

Chintan Tank, *Indiana University and General Sentiment*
Micah Linnemeier, *Indiana University and University of Michigan*
Chin Hua Kong and Katy Börner, *Indiana University*

Abstract

In support of researcher networking, VIVO provides temporal, geospatial, topical, and network analysis and visualization of data at the individual (micro), local (meso), and global (macro) levels. While a simple search might suffice to find a collaborator by name or keyword, VIVO visualizations make it possible to explore the data at a higher level of abstraction, using temporal trends, geospatial maps, expertise profiles, and scholarly networks. As VIVO is used across disciplinary, institutional, and national boundaries, the main VIVO visualizations must be domain independent and legible for a diverse set of users in a variety of countries. Given that different VIVO instances have very different data quality and coverage, and that they undergo many data changes during their creation and development, the visualizations must "degrade gracefully." For example, small numbers of records should not be visualized; by the same token, a paper with 800 authors should not be displayed as 800 fully connected nodes that, at best, resemble a "spaghetti ball." Data must be normalized carefully to ensure legibility of small and large numbers of records while also supporting comparisons.

Visualizations exist as part of the VIVO software itself or can be created using external tools and services that access VIVO data by querying its SPARQL endpoint. While external visualizations are typically created using data from one VIVO instance, there exist applications/services that analyze and display data from multiple VIVO instances as well as from other National Research Network instances (see Section 7.4.3). Visualizations can be static or interactive, offline or online, restricted access or freely available to anyone. A major goal of the VIVO visualization team at Indiana University was to empower others to run their own analyses of VIVO data and to create their own visualizations. Hence, this chapter also features insightful visualizations designed by others.

The structure of the chapter is as follows. Section 7.1 introduces the development goals and visualization design philosophy. Section 7.2 introduces the existing visualizations within VIVO. Section 7.3 describes the client-server visualization system architecture. Section 7.4 explains how to access, mine, and visualize VIVO data using different tools. Finally, Section 7.5 features insightful visualizations of VIVO and other national researcher network instances. We conclude with a discussion and outlook.

Keywords

visualization, data mining, Semantic Web data

7.1   VISUALIZATION DESIGN PHILOSOPHY AND DEVELOPMENT GOALS

7.1.1   USER FRIENDLY AND INFORMATIVE

As VIVO is used across disciplinary, institutional, and national boundaries, the main VIVO visualizations must be domain independent and legible by a diverse set of users in different countries. Some users might not have extensive time to learn how to read the visualizations. Hence, the decision was made to focus on implementing visualizations that are easy to read without much training. That is, instead of implementing highly sophisticated visualizations that use novel layout algorithms and unusual graphic design, VIVO visualizations use well-known reference systems such as timelines and simple color and size coding to represent data—at the individual to departmental to institutional level. A total of only four different reference systems are used: a timeline for temporal analysis; a map of the U.S. for geospatial analysis and visualizations; a map of all of science to represent expertise profiles; and egocentric network visualizations to render co-author and co-investigator networks. All four reference systems are deterministic and unchanging—i.e., users can learn these coordinate systems once, and different visualizations of the same type are easily comparable.

7.1.2   GRACEFULLY DEGRADING

Given that different VIVO instances have very different data quality and coverage, and that they undergo many data changes during their creation and development, the visualizations must "degrade gracefully." That is, small numbers of records should not be visualized; by the same principle, a paper with 800 authors should not be displayed as 800 fully connected nodes that, at best, resemble a "spaghetti ball." Data must be carefully normalized to ensure legibility of small and large numbers of records while also supporting comparisons. Hence, much effort was spent identifying and preventing visually awkward combinations of datasets, analysis, and visualization workflows. For example, a timeline is only rendered if at least two records are given; only papers with more than one author are used in the construction of co-author networks—all others are excluded but are listed and can be downloaded.

7.1.3   MODULAR AND ROBUST SOFTWARE

VIVO visualizations were implemented using the Model-View-Controller (MVC) paradigm: the model that encapsulates core application data and domain objects, the view that observes the state and presents output to the user, and the controller that receives the user requests from a view and performs operations on the model. The visualization model is stored in cache as RDF data and queried using SPARQL. The view is written in HTML and JavaScript and generated using the FreeMarker template language. The controller contains six components (see Figure 7.6 and discussion in Section 7.3.3). VIVO visualizations are accessed using diverse hardware systems—from desktop computers to iPads to phones—that run different web browsers. Early VIVO visualizations such as the egocentric network visualizations were designed in Flash; after the introduction and wide adoption of iPads (which lack Flash support), development efforts were changed to HTML and JavaScript to ensure visualizations display properly on major platforms.

7.1.4   EXTENDIBLE AND WELL-DOCUMENTED SOFTWARE

Visualizations exist as part of the VIVO software itself, or they can be created using external tools and services that access VIVO data by querying its SPARQL endpoint. While the external visualizations are typically created using data from a single VIVO instance, there exist applications/services that analyze and display data from multiple VIVO instances as well as other National Research Network instances. Visualizations might be static or interactive, offline or online, restricted access or freely available to anyone. A major goal of the VIVO visualization team at Indiana University was to empower others to run their own analyses of VIVO data and to create their own visualizations (see also Section 7.5 that features insightful visualizations designed by others). Consequently, major resources were spent on highly modular code that has well-defined APIs and is documented in detail. Tutorials were developed that empower others to create novel VIVO visualizations (see Section 7.5). Users might be interested to download and further analyze displayed datasets—e.g., by using Excel or network analysis toolkits. All visualizations allow users to download their underlying data, providing an easy-to-use point of access to various subsets of VIVO's triple-store data.

7.2   SOCIAL NETWORK VISUALIZATIONS

This section first presents the four visualizations—Sparkline, Temporal Graph, Map of Science, and Network Visualization—that are an integral part of the VIVO software, release 1.3 and higher.

7.2.1   SPARKLINE

A Sparkline is a small line chart that is typically drawn without axes or coordinates [1]. In VIVO, it is used to give a quick overview of a person's number of publications per year (see Figure 7.1) or the number of co-authors and publications (see Figure 7.5).

Figure 7.1: Sparkline visualization in top right (see also `http://vivo.ufl.eduto/display/n25562`).

7.2.2   TEMPORAL GRAPH

To identify and compare temporal trends of funding intake and publication output activity, Temporal Graph visualization was implemented (see Figure 7.2). The visualization allows users to visually and numerically compare the publication and grant activity of up to ten organizational units at different levels of detail—from schools within a university to individual researchers within a given department.

Publication and funding data can be compared by year in the line graph area of the visualizations (upper right), or in terms of their raw totals (bottom right). Organizational units are sorted in relation to their total publications or grants, but they can also be found by name using the search field. The underlying data for this visualization can be downloaded in CSV format.

7.2.3   MAP OF SCIENCE

VIVO implements the UCSD map of science and classification system, which was created by mining and analyzing millions of papers and their references from Elsevier's *Scopus* (2001-2008) and Thomson Reuters' *Web of Science (WoS) Science, Social Science, Arts & Humanities Citation Indexes* (2001-2010)—a total of about 25,000 source titles [2]. The resulting map shows the structure and

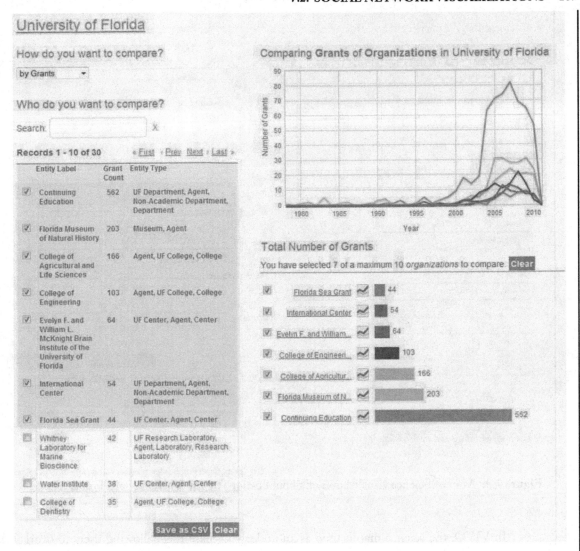

Figure 7.2: Temporal Graph visualization.

interrelations of 554 sub-disciplines of science in a spatial format. Each sub-discipline in the map of science represents a specific set of journals. Subdisciplines are further aggregated into 13 high-level disciplines based on natural visual groupings within the map, which are then assigned names and colors (see Figure 7.3).

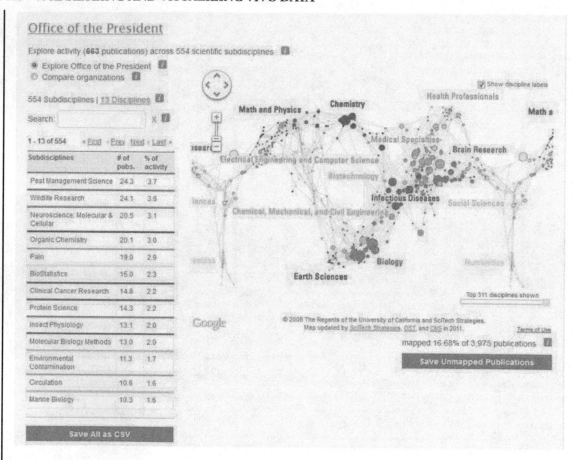

Figure 7.3: Map of Science visualization of all publications by the University of Florida.

In VIVO, the science map is used as an underlying base map, allowing users to overlay the publication-based expertise profiles of people, departments, schools, institutions, and other nodes in the organizational hierarchy.

The science map interface consists of a sortable table on the left and a map on the right that exhibits the same discipline color coding. Both are "tightly coupled"—for instance, selecting a (sub)discipline in the table leads to highlighting (via black outlines) of the respective nodes in the science map on the right. The table column labeled "# of pubs" shows how many of the publications were mapped to this (sub)discipline. The area size of nodes overlaid on the science map corresponds to that value. An equal distribution of papers over (sub)disciplines presents a balanced expertise profile, while a skewed distribution helps identify areas of specialization. A slider was provided to control the number of (sub)discipline overlays shown and to reduce visual clutter. Publication or counts may be fractional in some cases since some highly interdisciplinary journals are associated with more than one (sub)discipline. For example, the journal *Science* is fractionally associated with 82 subdisciplines. Starting with VIVO release 1.3, up to three maps of science can be compared; different colors help identify profiles of different organizations at the same hierarchical level—e.g., three researchers or two departments (see Figure 7.4).

7.2.4   NETWORK VISUALIZATION

Collaboration networks can be extracted from papers and grants, and are called co-author and co-investigator networks, respectively. As can be seen in Figure 7.5, networks are laid out in an egocentric manner: one researcher is placed in the middle and is surrounded by all of his/her collaborators to form a circular layout. Collaborator nodes can be laid out alphabetically or according to algorithmically calculated communities. Nodes are area-size- and color-coded by the number of collaborators; edges are width-coded by the number of times two people collaborated.

Users can traverse the collaboration networks locally: clicking on a collaborator node in the network brings up that person's profile on the left; selecting "Co-Author Network" right below a person's name renders his/her collaboration network. If funded grant awards were loaded, then a "Co-Investigator Network" icon and link can be seen in the top-right corner, and a click on it brings up this view.

7.3   VIVO VISUALIZATION SYSTEM ARCHITECTURE

7.3.1   FRONT-END VISUALIZATION LIBRARIES

VIVO visualizations use four existing visualization libraries:

1. Flot—JavaScript plotting library for jQuery.

   - Produces graphical plots of arbitrary datasets on-the-fly, client-side.
   - Works in modern browsers including IE and on iOS platforms.

2. DataTables[146]—JavaScript tabulating library for jQuery.

   - Client-side library for tabulating data with pagination.
   - Multi-column sorting with data-type detection.
   - Instant filtering of rows.

3. jQuery[147]—fast and concise JavaScript library.

   - Simplifies HTML document traversing, event handling, animating, and AJAX interactions.

[146]http://www.datatables.net
[147]http://jquery.com

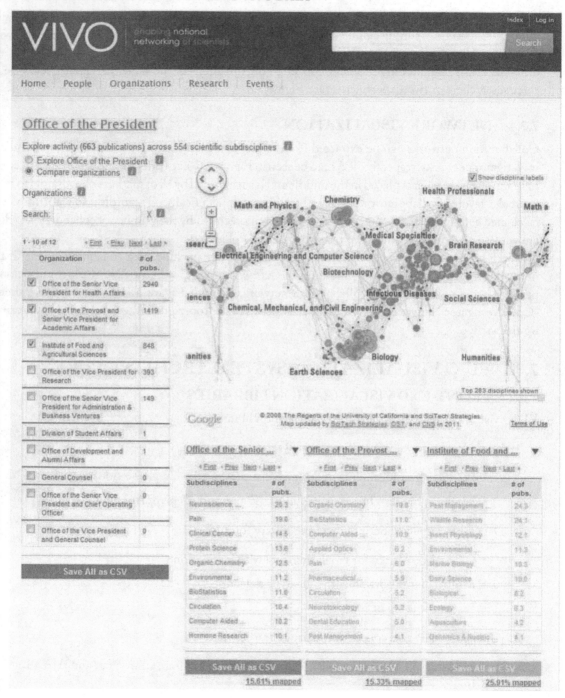

Figure 7.4: Map of Science visualization comparison of three different departments.

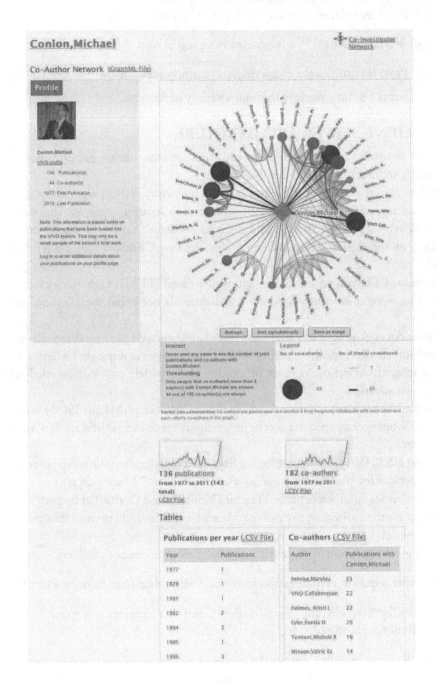

Figure 7.5: Egocentric network visualization of Michael Conlon's Co-Author Network.

- Provides uniform behavior on all major browsers.

4. Google Maps API v3[148]—JavaScript mapping library.

- Provides utilities for displaying and manipulating maps.
- Supports data overlays through a variety of services.

7.3.2   CLIENT-SERVER ARCHITECTURE

All but the network visualization follow the same software design pattern, using the client-server visualization architecture depicted in Figure 7.6.

Upon restart of the VIVO server, all different visualization algorithms are registered using Dependency Injection.[149] Four different visualization request types are supported in VIVO (explained below), each of which returns visualization content tailored for a different sort of context. Depending on the request type, a different "Visualization Controller" (see top middle of Figure 7.6) is called:

Standard. The response is a complete, well-formed HTML page that includes the visualization content. This is used in cases where the visualization is not being used "inline" and must stand on its own.

Ajax. An AJAX request is made and results are displayed as a visualization—e.g., a Sparkline or additional information such as a text or table. This response is requested whenever the visualization will be displayed within the context of another page, which may not be solely dedicated to this particular visualization.

Data. The response is a downloadable file in CSV, GraphML, or JSON format. This response is requested whenever a particular user or program wishes to access the data that underlies a particular visualization.

Short URL. Wrapper request for the **Standard** visualization. This type of request has a pattern: "/vis/< short key for the visualization >/< optional uri >." Once this type of request is made, the controller provides parameters to the Standard Visualization Controller by parsing the request URL. This request method allows us to display and exchange more human-readable and attractive versions of visualization URLs.

The general process of requesting a visualization proceeds as follows.

1. The user requests a visualization from the 'Client-side' (e.g., using a short URL).

2. On the 'Server-side,' the 'Visualization Controller' receives the request and gives control to the 'Request Handler.'

3. The 'Request Handler' checks the requesting user's privileges against the privileges required by the visualization and, if OK, delegates control to the particular visualization requested by

---

[148]http://code.google.com/apis/maps/documentation/javascript
[149]http://en.wikipedia.org/wiki/Dependency_injection

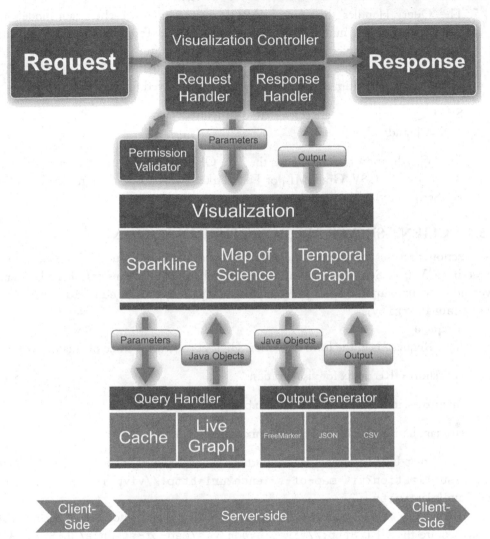

Figure 7.6: Client-server visualization architecture.

the user. If the permissions are insufficient or the requested visualization does not exist, then it responds with an error message. This mechanism allows for the creation of visualizations intended only for a subset of users (administrators at a particular university, for example).

4. Once the 'Visualization Handler' receives a request, it verifies all the required parameters and passes them on to the 'Query Handler.'

5. The 'Query Handler' then runs SPARQL queries against cached data models for specific visualizations—e.g., number of papers per year for the 'Temporal Graph' visualization. Visualizations might depict large amounts of data but running the query against the "live" data model is both memory-intensive and time-consuming. Efficient caching of SPARQL queries speeds up response times; plus, query results can be saved for later reuse.

6. SPARQL query results are converted to structured 'Java Objects' and returned to the 'Visualization Handler.'

7. The 'Visualization Handler' sends the 'Java Objects' to an 'Output Generator' that converts it into JSON, CSV, GraphML, or FreeMarker (which is a combination of HTML, CSS, JavaScript, etc.) in the user's browser.

7.3.3   CLIENT-SERVER ARCHITECTURE IN ACTION

This section discusses the usage of the architecture presented in Section 7.3.2 when requesting and rendering a 'Map of Science' visualization. Specifically, all publications at Indiana University will be overlaid over the map of science (see sample at `http://vivo.iu.edu/vis/map-of-science/IndianaUniversity`).

Request
The 'Request' is in the form of a 'Short URL' consisting of three mandatory parameters:

- /vis: short URL prefix for visualization

- /map-of-science: type of visualization being invoked

- /IndianaUniversity: level of the organizational hierarchy

A 'Long-URL' format is supported as well and looks like `http://vivo.iu.edu/visualization?vis=map-of-science&uri=http://vivo.iu.edu/individual/IndianaUniversity`.

Requesting this visualization for School or Library and Information Science (SLIS) data would require the URL `http://vivo.iu.edu/vis/map-of-science/BL-SLIS`. A co-author network for a researcher with ID 25557 can be requested using `http://vivo.iu.edu/vis/author-network/person25557`. Note that the URI is optional (with the 'Temporal Graph' visualization, for example). In such cases, it is the responsibility of the visualization developer to use a reasonable default value—e.g., the root node of the organizational hierarchy.

Request Handler
All new visualizations need to implement the interface VisualizationRequestHandler to ensure that requests for different visualization types can be handled and user access rights can be validated. The latter access right control was introduced in VIVO 1.3. While all visualizations are viewable by all users currently, the getRequiredPrivileges function does support privilege checking in support

of future use cases where specific visualizations might only be accessible to users with appropriate privileges.

Response Handler

As the name suggests, the 'Response Handler' takes the output provided by the visualization and forwards it to the client-side. It makes sure that different types of outputs like FreeMarker response, JSON, and CSV are forwarded with appropriate HEADERS so that the browser is able to interpret it properly. If a 'Standard' visualization request is made, it compiles the complete HTML markup from the FreeMarker visualization template and other variables.

Visualization Handler

The 'Visualization Handler' processes the request (in the form of parameters) and responds with a visualization output. As shown in Figure 7.6, it makes use of two sub-components, the 'Query Handler' and 'Output Generator,' to serve the request.

Query Handler

This handler is responsible for generating plain-old Java objects (POJOs) from the results of a SPARQL query fired against the data backend. The SPARQL query is formed by filling in the blanks in the query template with the parameters passed. A template exists illustrating how the SPARQL query should be formed. The type of visualization output depends on the visualization in question. Starting with the VIVO 1.3 release, the 'Query Handler' fires SPARQL queries against cached models for certain classes of visualizations like 'Temporal Graph' and 'Map of Science' instead of firing them against the live data model.

Cache

Visualizations like the 'Temporal Graph' or the 'Map of Science' involve calculating total counts of publications or grants for some entity. These counts are stored with people nodes. Retrieving counts for higher-level entities—e.g., a department or entire university—requires traversal of the complete organization hierarchy below an entity—e.g., all colleagues, schools, departments, and their researchers—to retrieve and sum up publication and funding records. These queries are memory-intensive and time-consuming; hence, a caching solution was implemented to precompute and store information about the number of papers/grants for each node in the organizational hierarchy. The values are computed on the first user request after a server restart and are stored in the RDF model. Future releases will store data models on disk and update them periodically.

Live Graph

The 'Query Handler' also supports firing SPARQL queries against non-cached VIVO data structures.

Query Result Format

Query results are returned as value objects (i.e., placeholders for structured data to be referenced later) that detail:

- Entity (for subject entity)

- SubEntity (for child entities)

- Activity (for publication info)

Output Generator

The 'Output Generator' is responsible for generating the visualization output from the POJOs[150] returned by the 'Query Handler.'

Visualization Controller

As discussed in the beginning of this section, the 'Visualization Controller' (see top middle of Figure 7.6) supports four different types of visualization: Standard, Ajax, Data, and Short URL. The respective controllers expect different types of data input:

Standard. Requires ResponseValues that are understood by the underlying FreeMarker templating system.

Ajax. Reads TemplateResponseValues that are also understood by the FreeMarker templating system but which are directly written to the HttpServletResponse stream.

Data. Requires a map consisting of header information and the actual content of the data, directly written into the HttpServletResponse stream.

Short URL. Same as Standard output format.

A 'Map of Science' visualization is generated via a FreeMarker object that renders a web page using HTML markup including CSS, JavaScript, JSON (used by JavaScript to render tables, maps, etc.), and CSV (for downloading data from the tables).

Response

On the client-side, the user either sees a web page containing the 'Map of Science' (in case of "standalone" render_mode) or a prompt to save the CSV (in case of "data" render_mode).

7.4 ACCESSING, MINING, AND VISUALIZING VIVO DATA

VIVO data can be accessed, mined, and visualized by (1) saving precompiled data from VIVO visualization interfaces; (2) using visualization design templates to implement novel VIVO visualizations; or (3) running SPARQL queries against VIVO RDF triple stores.

---

[150]http://en.wikipedia.org/wiki/Plain_Old_Java_Object

7.4.1   DATA PROVIDED BY VIVO VISUALIZATIONS

All but the Sparkline visualization interface provide easy access to precompiled VIVO data.

Temporal Graph supports download of entity name, type and publication/grant count in CSV format.

Map of Science visualization supports download of the fractionally assigned number of publications per (sub)discipline.

Network Visualization lets users download CSV files of publications per year or all co-authors as CSV files plus the complete collaboration graph as a GraphML file for further processing. For example, external tools like Network Workbench (NWB) [3] or Gephi [4] can be applied to re-render Michael Conlon's network from Figure 7.5. In Figure 7.7, the left image (a) shows the complete network, while the right network (b) shows the remaining structure after Conlon's node was deleted.

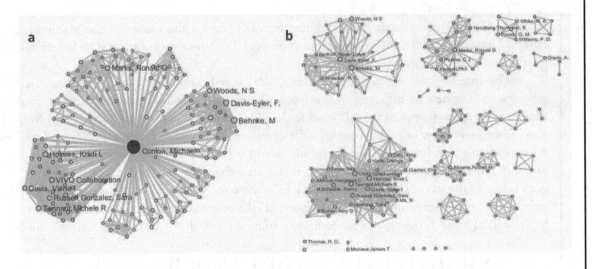

Figure 7.7: Michael Conlon's Co-Author Network re-rendered in NWB (left) and without Conlon's node (right).

7.4.2   DATA ACCESS AND VISUALIZATION USING VISUALIZATION TEMPLATES

The creation of a new VIVO visualization requires the following:

- identification of the data to be represented;

- identification of the visual representation;

- creation of SPARQL queries to retrieve the needed data;

- implementation of a back-end module for a new visualization;

- implementation of a front-end module that renders the data created by the back-end; and

- connection of back-end and front-end to get a complete working pipeline.

The 'VIVO Visualization Tutorial'[151] (slides 102+) details these steps for the design of a word cloud visualization that shows all words from publication titles for a selected person, size-coded by number of occurrence.

7.4.3   DATA RETRIEVAL VIA SPARQL QUERIES OR DUMPS

Using a VIVO SPARQL query endpoint such as the one provided by the University of Florida (UFL) at `http://sparql.vivo.ufl.edu/sparql.html`, RDF queries can be run by anyone. The results can be saved in JSON, text, CSV, or TSV format. See the 'VIVO Visualization Tutorial'[152] (slides 42–65) for more details.

As SPARQL queries are resource-intensive, VIVO data is also made available via simple data dumps. These data dumps can be loaded into an empty VIVO instance or a triple store, where custom queries can be run against them.

The International Researcher Network (IRN) visualization at `http://nrn.cns.iu.edu` provides an overview of different system types—Elsevier's Collexis, Harvard's Catalyst Profiles, Stanford's CAPS system, and the NIH-funded VIVO system—that are in production worldwide. It also shows the number of loaded people profiles, publications, patents, funding awards, and courses. System adoption and data loading can be animated for January 2010 to today, with data updated monthly. Hover the mouse over a data point that is area-size-coded to see data details; click on it to get a listing of the records below the map area; hide/show the map as desired. It is our hope that this website will encourage future adoption and usage of IRN systems in support of scientific discoveries, technological breakthroughs, and the communication of research results to diverse stakeholders.

7.5   INSIGHTFUL VISUALIZATIONS OF IRN DATA

VIVO data and data from other international researcher networking sites can be used to provide insight into expertise at a university on a certain topic, win major center funding, or help study and utilize cross-institutional expertise networks. Selected data analyses by researchers outside of the VIVO team are discussed below.

7.5.1   COLLABORATION PATTERNS FOR MEDICAL INSTITUTIONS

Simon Porter at Melbourne University used data from a local VIVO instance to extract a bimodal network of search terms and researchers from research profile search results to show the university's

---

[151]http://ivl.cns.iu.edu/km/pres/2011-borner-VIVO-Workshop.pdf
[152]http://ivl.cns.iu.edu/km/pres/2011-borner-VIVO-Workshop.pdf

Figure 7.8: Evolution of the International Researcher Network.

capability in "Disaster Management" to government officials (see Figure 7.9). Here, search terms are represented by black squares, while researchers are denoted by circles of different colors. Node colors indicate the School, and node size represents the number of matching search terms. Researcher names have been removed as this work is ongoing.

7.5.2   TOP MESH DISEASE CONCEPTS APPEARING IN PUBMED PUBLICATIONS

Jeffrey Horon, formerly at the University of Michigan, now with Elsevier, Inc., visualized the Medical Subject Headings (MeSH) associated with publications by researchers at the University of Michigan Medical School (see Figure 7.10). In Horon's visualization, links connect concepts in cases where 100+ authors published about the same 2 concepts within the span of their careers. This visualization

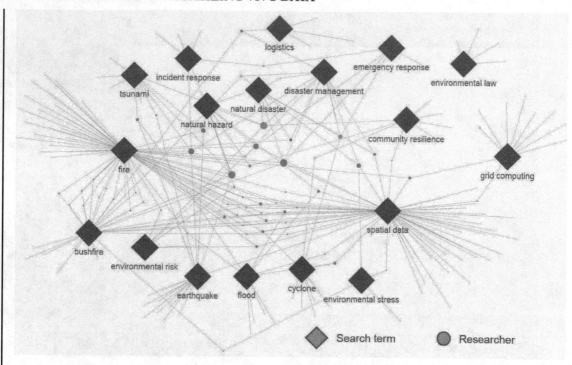

Figure 7.9: Bimodal network of search terms and researcher expertise.

revealed that "animal disease models" were central to disease research, which encouraged additional attention to animal husbandry, animal expenses, and core/shared services overall.

7.5.3   IDENTIFICATION OF COLLABORATION NETWORKS IN SUPPORT OF FUNDING PROPOSALS

Jeffrey Horon also mapped the co-participation and co-authorship networks of researchers at the University of Michigan Medical School in preparation of a P30 NIH Center grant application (see Figure 7.11). The initial networks were used to identify relevant expertise at the university; a visualization of existing collaboration linked among members of the final project team was included in the successful grant application. Shown below are the PI's relationships with various P30 members, conveying that the PI was not only the formal center of the group but also the informal center and the person who exhibited the highest "betweenness centrality."

7.5.4   INTER-INSTITUTIONAL COLLABORATION EXPLORER

Nick Benik and Griffin Weber at the Harvard Medical School in Boston developed a novel means to access "collaborative publications" found at two or more researcher networking websites (see

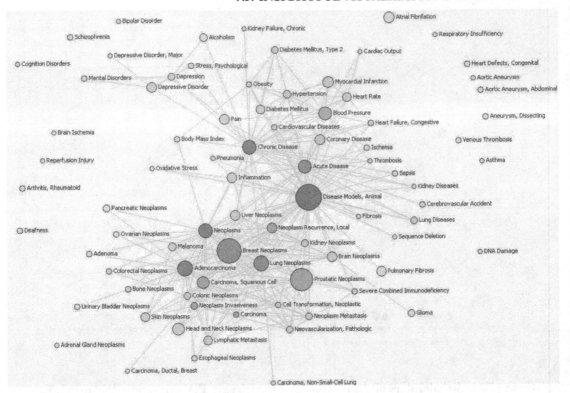

Figure 7.10: Top MeSH disease concepts appearing in PubMed publications by the University of Michigan Medical School.

Figure 7.12). The idea that institutions don't work together and that biomedical research is conducted in silos is not true. Researchers, even when separated by great distances, are indeed willing to work together, and this visualization demonstrates that they often do. The interactive visualization can be explored at http://xcite.hackerceo.org/VIVOviz.

The explorer shows collaboration patterns between the 11 color-coded U.S. universities listed on the top left. The number on the right of each institution indicates the number of publications (e.g., 15.124 for Harvard Medical School). The visualization on the right organizes the 11 institutions in a circular fashion and assigns each an arc which is proportional to the number of collaborative publications found on the site. The inner colored bands represent the number of collaborative publications found between the two institutions that each band connects. Clicking an institution's arc will hide any bands not connected to that institution and will display a timeline of when that institution's collaborative publications were written. See also the lower part of the figure that shows a zoom into the collaboration circle when the Northwestern Medical School is selected.

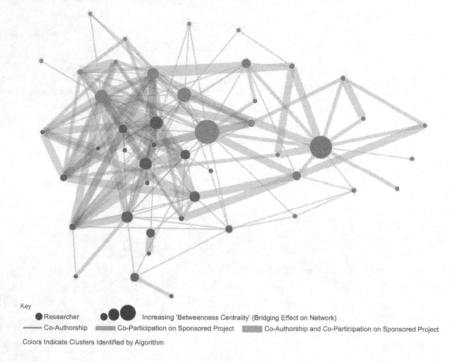

Key
● Researcher   ● ● ● Increasing 'Betweenness Centrality' (Bridging Effect on Network)
── Co-Authorship   ▨ Co-Participation on Sponsored Project   ▨ Co-Authorship and Co-Participation on Sponsored Project
Colors Indicate Clusters Identified by Algorithm

Figure 7.11: Relationships of the P30 principal investigator (PI) with various P30 members.

7.6   DISCUSSION AND OUTLOOK

7.6.1   OPEN SOCIAL CONTAINERS AND GADGETS

A key development goal for VIVO is providing easy access to VIVO data and creating highly modular and extensible code. Ideally, a "million minds" can benefit from VIVO and/or add value to VIVO data and code. It is desirable to have an API that makes it easy for VIVO and other developers to design new views of the data. OpenSocial (http://code.google.com/apis/opensocial) defines a common API for social applications across multiple websites. That is, if VIVO would implement the API, it could host third-party applications or so-called OpenSocial gadgets. Sample gadgets might be applications that calculate the h-index [5] for a person, link to relevant data sets on Data.gov, display events from a calendar of upcoming presentations, or perform many of the other functions listed at http://www.applications.sciverse.com/action/gallery. VIVO users could select a certain set of gadgets and arrange them on their personal VIVO profile page (e.g., stacked vertically on the right) as a means to customize their profile. Institutions might like to implement novel gadgets that highlight specific features of their activities (e.g., extensive collaborations with local companies). Applications that support OpenSocial (i.e., provide an OpenSocial container) also make data access easier. In addition, VIVO visualizations that are OpenSocial-compliant (i.e., are OpenSocial gadgets)

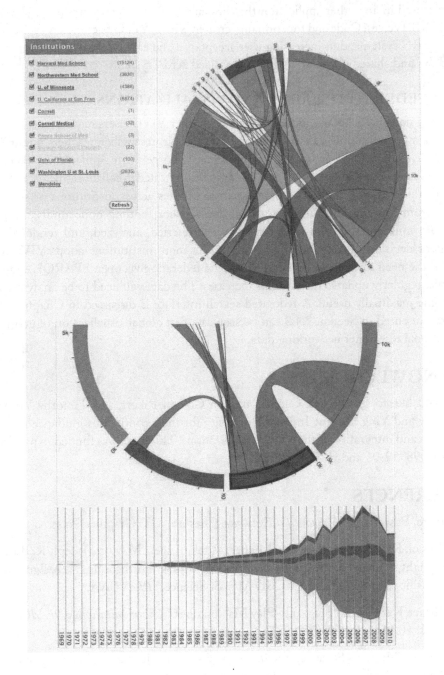

Figure 7.12: Inter-institutional Collaboration Explorer.

can be reused in any other application that has an OpenSocial container (e.g., Harvard Profiles or Elsevier's SciVerse). Code and functionality created within VIVO becomes available across different types of IRN systems, thus increasing user acceptance and ease of use due to common data analysis workflows and shared visual language standardized across systems.

7.6.2   FEDERATED SEARCH AND VISUALIZATIONS

The many needs of scholars, administrators, and other VIVO users require queries and analysis of VIVO data from multiple VIVO instances. For example, researcher-teachers might like to identify those colleagues in their geospatial neighborhood who perform similar/complementary research, teach similar or complementary courses, or might be interested in collaborating on a funding proposal. Administrators might like to compare expertise profiles across institutions or synergistically pull experts from multiple institutions to apply for a major funding opportunity. In all these cases, data from multiple VIVO instances needs to be queried, analyzed, and rendered in a way that intelligent decision-making becomes possible. As more institutions adopt VIVO or VIVO-like systems, the need for cross-search and cross-data federation via open SPARQL endpoints or freely available, regularly updated data dumps increases. The data will need to be analyzed and visualized to become maximally useful. A federated search interface is discussed in Chapter 6, and the IRN website presented in Section 7.4.3 can be seen as a first global visualization of geospatially located international researcher networking data.

ACKNOWLEDGMENTS

We would like to thank Nick Cappadona, Jon Corson-Rikert, and Timothy Worrall at Cornell University and Ying Ding at Indiana University for their continuous guidance during the implementation and integration of the VIVO visualizations. This work was funded in part by NIH awards U01GM098959-01 and U24RR029822.

REFERENCES

[1] Tufte, Edward. 2006. *Beautiful Evidence*. Cheshire, CT: Graphics Press.

[2] Börner, Katy, Richard Klavans, Michael Patek, Angela M. Zoss, Joseph R. Biberstine, Robert P. Light, Vincent Larivière, and Kevin W. Boyack. Forthcoming. "Design and Update of a Classification System: The UCSD Map of Science." *PLoS ONE*.

[3] Börner, Katy. 2011. "Plug-and-Play Macroscopes." *Communications of the ACM* 54 (3): 60–69. DOI: 10.1145/1897852.1897871

[4] Bastian, Mathieu, Sebastien Heymann, and Mathieu Jacomy. 2009. "Gephi: An Open Source Software for Exploring and Manipulating Networks." In *Proceedings of the Third International AAAI Conference on Weblogs and Social Media, San Jose, California, May 17–20*. AAAI Publications.

[5] Hirsch, J. E. 2005. "An Index to Quantify an Individual's Scientific Research Output. *PNAS* 102 (46): 16569–16572. DOI: 10.1007/s11192-010-0193-9

CHAPTER 8

The Future of VIVO: Growing the Community

Dean B. Krafft, *Cornell University Library*
Katy Börner, *Indiana University*
Jon Corson-Rikert, *Cornell University*
Kristi L. Holmes, *Washington University*

Abstract

The VIVO effort is growing beyond single institutions, beyond the initial implementors, and beyond the VIVO software platform, to become a standard for exchanging information about researchers and research. As part of this process, the open-source development effort is moving to a new community framework, and the VIVO ontology is being adopted by commercial, governmental, and consortial organizations.

Keywords

software development, open source, community source, linked open data, Semantic Web, ontology

8.1 INTRODUCTION

The VIVO approach to scholarly networking combines local institutional control with the interoperability afforded by linked open data. This is not an accidental pairing—to be successful, VIVO needs grounding in the local institution for sustainability and authoritativeness. At the same time, extending the reach of university research beyond the campus has become an important component in establishing and maintaining a competitive research program.

National and international efforts to coordinate research also depend on access to high-quality, reliable data. A number of efforts to create centralized systems have not proven sustainable due to the high rate of change and lack of incentives for individual researchers and institutions to participate. Several developments in recent years show promise for changing both the incentives and the outcomes.

There has been a significant shift in research funding programs to favor and even require collaborations across institutions and disciplines. Changes in the nature and practice of science, perhaps most evident in the life sciences, have also increased the need to represent more disciplines on larger projects—not just because research is becoming more specialized, but also because more specialized skills are relevant. As the benefits of a team-based approach to research and scholarship become more widely accepted, computer science departments are hiring biologists, physicists are learning biology, and medical journals are in high demand in engineering schools as even applied scientists engage in what was once purely biomedical research.

Data interchange and interoperability have been critical IT requirements for decades. Linked open data[153] relies on a very simple common data model to support a very open-ended path for data exchange where interoperability becomes dependent only on agreement on meaning, bypassing time-consuming efforts at data model alignment and format conversion. Linked open data standards are maturing and tools are evolving in the commercial as well as open-source realm to meet the challenges of scale, performance, and ease of use. The once near-vertical learning curve for Semantic Web applications is flattening, with a much healthier set of options for ontology editing, data conversion to RDF, and SPARQL query support.

VIVO has been focused on data sharing, reuse, and redeployment in new applications from the beginning. The VIVO system provides very usable and accessible tools for editing ontologies, editing and managing RDF statements, and making all of it available as both web pages and linked open data. Now that we have data and we have tools, the vision of what to build and where to go next is beginning to take tangible form. We are ready to grow both the community using VIVO to discover and exchange information and also the depth and capabilities of the tools that make use of VIVO linked open data.

8.2   UPCOMING RESEARCH AND DEVELOPMENT

Based on our experience with VIVO to date and input from VIVO developers, implementers, and related profiling projects, we find that there are a clear set of short-term research and development tasks. These fall into three general areas: developing the core VIVO application (covered in Section 8.2.1 below); supporting VIVO collaboration and discovery networks (covered in Section 8.2.2 below); and evolving the VIVO ontology (covered in Chapter 2). As we move beyond the clear deliverables of the NIH-funded VIVO project, one of our major challenges will be identifying, prioritizing, and allocating the research and development tasks to address the future of VIVO. We will discuss that community process in Section 8.5 below. In this section, we will focus on some of the immediate research and development priorities.

8.2.1   DEVELOPING THE VIVO APPLICATION

As this is being written, the specific features and development priorities for the next VIVO release are being finalized. This is happening in the context of longer-term development goals that focus on scalability, modularity, improved site management and reporting, network features, and the intro-

---

[153]http://linkeddata.org

duction of individual-level controls over presentation and data sharing. These goals have emerged from the intensive three-year development period of the NIH VIVO grant and reflect the need to make the VIVO code base transparent, extensible, and responsive to requests from users and adopting institutions.

It is important to note the unique features that will remain central to VIVO. The VIVO application is ontology-driven rather than relying on a single, fixed database schema for its content; this enables the VIVO application to respond gracefully to changes in the ontology and to evolve, potentially radically, in response to the local needs of adopting institutions. VIVO will continue to improve its ontology-editing and content-editing tools along with new public-facing features. Users have confirmed that VIVO's ability to model, create, and display content through a single application allows for rapid implementation of VIVO and helps to reduce the learning curve for new staff by providing immediate feedback on additions and changes.

Scalability

VIVO has relied on the Jena Semantic Web Framework since 2007 for its code libraries and data persistence in a triple store. The VIVO 1.2 release in February 2011 featured conversion from Jena's in-memory RDB triple store to the database-hosted SDB triple store. While SDB is much more scalable, through reliance primarily on disk storage rather than available memory, the rapid evolution of triple-store technologies has motivated development to make VIVO independent of any single triple-store framework. By converting VIVO's data access and storage interface to rely only on the SPARQL query standard, VIVO with version 1.5 (scheduled for summer 2012) will be capable of using any triple store supporting SPARQL for both query and update. This change will enable implementing sites to choose the triple store that provides the best blend of scalability, query, and update response to meet their needs.

Modularity

Developers first encountering VIVO have often suggested that aligning VIVO with a well-known Java framework would speed their initial orientation to the VIVO code base and facilitate building extensions to VIVO. The VIVO development team has committed to moving VIVO to the OSGi framework over the next several releases. OSGi provides a service-oriented architecture approach and a flexible degree of modularity that fits VIVO's current structure and will provide both direction and transparency as key components are rearchitected in coming releases.

Site Management and Reporting

As universities, government agencies, professional societies, and other types of institutions and organizations adopt VIVO, the number and nature of functions required for managing a VIVO instance have steadily increased. The investment in acquiring and disambiguating data for VIVO is often significant, and each VIVO installation must provide a very tangible return on that investment, including queries and reports. VIVO 1.5 will introduce two new types of VIVO pages: simple HTML pages for presenting contextual information about an institution, and SPARQL query pages for supporting reports. Query pages will support one or more site-specific, parameterized queries to highlight data such as recent grants or publications at a site. These simple reports will provide an open-ended way to showcase VIVO content in a manner that is helpful for institutional workflows and reporting. Reporting capabilities are expected to grow in sophistication in future releases as effective report specifications are developed and shared among the VIVO community.

In another area of reporting, the University of Florida recently introduced logging features in VIVO that permit tracking additions and deletions to VIVO data by date, time, and user agent, whether introduced through the VIVO Harvester or by interactive editing in VIVO itself. These changes will be folded into the main VIVO code for optional activation at other institutions according to local needs.

Network Features

The VIVO application highlights only the most obvious connections inherent in VIVO data through bi-directional linking and navigation. While VIVO's visualization features bring a person's or department's relation to others very much to the forefront, many other aspects of VIVO data are not obvious in the current out-of-the-box interface.

VIVO's visual designers on the UI and visualization teams have proposed significant changes to VIVO's default displays to incorporate more network-centric features such as counts of co-authors and co-investigators, and to provide additional cues leading to existing visualizations. By highlighting the collaborative and connected nature of researchers" activities individually and in aggregation, VIVO brings a unique perspective to the web presence of an institution.

Individual Controls

As an almost entirely data-driven application, VIVO has heretofore offered very little customization at the individual page level, concentrating instead on configurations across all content by type (class) or relationship (property). With the VIVO 1.5 release, the application's primary data ontology and its existing user-accounts data graph and menu-page data graph will be supplemented by an application ontology affecting display and editing behavior in fundamental ways. This ontology provides a path for more individualized control over content within the application for VIVO site webmasters and, in coming releases, researchers themselves.

8.2.2   SUPPORTING VIVO COLLABORATION AND DISCOVERY NETWORKS

There is no doubt that successful researcher networking requires the ability to search across VIVO sites. The VIVO proposal was predicated on making linked data openly available beyond the home institution, and search is a natural first application to support, as one of the first steps in growing VIVO is extending its capabilities beyond a single institution.

Planning discussions started very early in the project, and a white paper outlining a plan for national search became the first section of a more comprehensive development roadmap prepared by the three development team leads at Cornell in the spring of 2010. Unlike the eagle-i project's federated search approach, the VIVO developers proposed creating one or more search indexes combining information from multiple VIVO instances. The contents of the index itself would not be exposed directly, avoiding concerns about sharing data outside the direct purview of the host institution; these concerns were rendered largely moot by the ready availability of VIVO data from each participating institution on the web via linked open data requests, although several institutions have policy concerns about sharing VIVO data in bulk.

As plans to develop a working, integrated VIVO-search prototype for the 2011 VIVO Conference materialized, the development team created a system that harvested data for a common search site using standard linked data requests and a separate software application from VIVO itself. By taking this approach, any institution or consortium may elect to create an index from any source of RDF expressed using the VIVO ontology, whether from VIVO, VIVO-ontology-compliant applications, simple directories of RDF files, or other software.

While the prototype search site at http://vivosearch.org currently covers only the seven VIVO project partners and Harvard University, the inclusion of Harvard is a notable demonstration of interoperability. The data from Harvard was not produced with the VIVO system, but it is being delivered using the standard VIVO ontology. The developers of Harvard Profiles elected in early 2011 to convert their underlying data to RDF using the VIVO ontology as the core internal representation.[154] Other researcher-profiling systems are now adopting the VIVO ontology as the standard interchange format for linked open data about research and researchers.[155]

This simple proof of concept illustrates key principles of a sustainable approach: data remain authoritative in the local VIVO instance and are indexed centrally to provide a responsive search experience for the user. The index can be updated in parallel at any desired frequency determined by the time to index each site; we anticipate that a daily frequency with overnight updates would normally be adequate. The Solr index includes the full text of each VIVO page but is structured to reflect the types and properties of the VIVO ontology; local detail is included in the index for completeness, but faceting is done at the VIVO core-ontology level to assure a consistent search experience. Users are directed back to local VIVO or Profiles institutional instances from the search results to provide appropriate branding and context.

While the idea of a single national or international search for researchers across multiple institutions and disciplines is quite appealing in the abstract, a number of practical and policy issues come immediately to mind: the classes or levels of institutions that should be included or excluded; whether and how to include government, private sector, and even unaffiliated researchers; how to govern and fund such a service; and whether existing technologies can expand to the scale required.

---

[154]http://profiles.catalyst.harvard.edu
[155]https://www.icts.uiowa.edu/confluence/display/ICTSit/Loki-Vivo+Alignment

Figure 8.1: The VIVO beta search website.

Assumptions will be called into question, including any notion of limiting participation and discovery along national boundaries.

For these and other practical reasons, integrated search services are more likely to be mounted and sustained across targeted consortia than at the national or international level, where issues of scale become large enough to require infrastructure at the level of a major commercial search engine. The 60 NIH Clinical and Translational Science Awards are one likely consortium; the principal investigators of the CTSAs voted overwhelmingly in October 2011 to "encourage their institution(s) to implement research networking tool(s) institution-wide that utilize RDF triples and an ontology compatible with the VIVO ontology."[156] Networks focused by geography, discipline, or

[156]https://www.ctsacentral.org/recommendations-and-best-practices-research-networking

funding source have all the advantages of coherence and avoid many potential scaling and governance pitfalls.

There is no reason a single VIVO institution cannot participate in several search networks simultaneously. The Weill Cornell Medical College in New York City is an urban campus affiliated with Cornell University, based four hours away in Ithaca, New York. Each campus maintains an independent VIVO instance but will be sharing a common search index to allow cross-campus discovery of research activities and expertise. Weill Cornell is also the lead institution in the New York City-based Clinical and Translational Research Center (CTSC),[157] an affiliation involving several hospitals and institutions of higher education independent from Cornell. The same VIVO data from Weill Cornell can be incorporated into the CTSC website and indexed for a CTSC-scoped search to meet that consortium's needs, leveraging the initial VIVO investment again while concentrating update and maintenance in a single local system.

VIVO data can also feed other networks and/or network reporting or analysis tools. The Federal Demonstration Partnership,[158] an association of federal agencies and academic research institutions, has begun piloting data acquisition and modeling for a federal Science Experts Network Curriculum Vitae, SciENCV,[159] using data in the VIVO ontology format from the University of Florida's VIVO system and Harvard Profiles, among other sources. Principles for the effort include open participation, sharing of data for public consumption as well as government use, and ownership by individuals of profile data about them.

Scaling this and other network efforts will require the development of new tools to address compliance testing for source data. Data disambiguation and quality control must be able to rely almost exclusively on automated techniques; these techniques will be far easier to implement if systems providing stable, unique identifiers for multiple types of data can be established and sustained at world scale. Looking first at identifiers for people, Open Research and Contributor ID (ORCID),[160] a non-profit organization chartered to enable reliable attribution of authors and contributors to published works, offers a promising model for addressing unique identifiers on an international scale, but it still needs to demonstrate a viable long-term business model.

It will be important for the researcher-networking community to demonstrate the benefits as well as the many challenges of coordination at a large scale. Visualizations provide new insights and highlight data anomalies; even simple inter-institutional searches can provide access to information that would be extraordinarily difficult to obtain without common VIVO data.

---

[157]http://weill.cornell.edu/ctsc/
[158]http://sites.nationalacademies.org/PGA/fdp/index.htm
[159]http://rbm.nih.gov/profile_project.htm
[160]http://about.orcid.org/

8.3   INTEGRATING VIVO INTO THE RESEARCHER ECOSYSTEM

While the adoption of VIVO at an institution is typically first motivated by organizational needs—whether centrally or in one or more colleges or schools at a larger institution—VIVO is unlikely to achieve long-term sustainability at that institution if individual researchers do not have confidence in their own representations and if they do not find the VIVO data or system useful.

The challenges of confidence can be addressed through quality-improvement processes looking at source data, conversion and update processes, and ways to lower barriers for data review and correction, as discussed in the two case studies in chapters four and five. Key elements of VIVO's utility have also been described in earlier chapters, and many of these benefits apply equally to individual researchers. There is a unique dynamic in how researchers adopt and apply tools in the course of their work, however. Seeking to understand this dynamic can hopefully help shape VIVO's evolution in ways that enhance its ability to facilitate research itself—to improve awareness of research beyond a single department or institution and to stimulate new ideas in any discipline.

This is a tall order. The value proposition for researchers too often focuses on minimizing the time and effort required to provide and review data for administrative purposes and grants administration. While important and valuable, other benefits can be coupled more closely to scholarly and research activities including mentoring, recruitment and orientation of new faculty and graduate students and ready access to recent work by colleagues outside one's own immediate circles.

We are also seeking to extend VIVO to support information about researchers and research that is not easily available through a Google search. This is happening in three areas: research resources and facilities; expertise, techniques, and certifications; and research dataset descriptions.

Members of the VIVO team are working together with the team that developed the eagle-i ontology describing research resources on a new CTSAconnect project. This will create an extended and combined ontology that will support directly relating researchers to the research resources and facilities they use. This, in turn, will allow researchers to discover potential collaborators based on the facilities or research resources (cell lines, analysis tools, etc.) with which they work.

Moving further in this direction, the VIVO ontology could be extended to represent specific certifications or expertise in specific research techniques. Often a researcher seeks to find a collaborator who can bring highly specific research skills to a project. VIVO can potentially support this kind of discovery across large, multi-institutional collaborations.

The third area, describing research datasets, has already seen significant work. The University of Melbourne, as part of the Australian National Data Service (ANDS) project, has extended the VIVO ontology to support research dataset description. The Institute for Museum and Library Sciences (IMLS) has funded Cornell University to take its previous work on the DataSTaR project[161] and use it to create a modular extension to VIVO to support dataset description.[162] Currently,

---

[161]http://datastar.mannlib.cornell.edu
[162]http://sites.google.com/site/datastarsite/

researchers have no standard way to easily describe and make available information about their research datasets. This will be a valuable new capability for the VIVO system.

Coupling VIVO data with an institutional repository (such as DSpace or Digital Commons) dovetails well with open-access policies adopted in recent years by universities such as MIT,[163] Harvard,[164] and University College London.[165] The University of Rochester has featured researcher pages and photos as a browse facet for its UR Research repository,[166] and similar coupling with VIVO could help create incentives for researchers to deposit publications and research data.

8.4   ENCOURAGING ADOPTION

Adoption comes in two forms: an institution can either adopt the full VIVO framework, or it can join the VIVO community by running a system that makes researcher and research information available as linked open data using the VIVO ontology. Either approach requires a significant investment of resources and high- level approvals within the institution. At a typical university, adoption of VIVO may require approval from the provost, the vice-provost for research, the CIO, and the university librarian. The data stewards must approve the use of public data from the human resources, registrar, grants, and other institutional databases. Last, but far from least, the university faculty must see VIVO as making their lives easier, rather than more difficult. As Chapter 3 makes clear, a local implementation of VIVO requires significant and ongoing investments in data ingest and update as well as a support effort for the local users of the VIVO platform to help them benefit from the rich structured data within the VIVO.

The benefits of running VIVO within a single institution can be very significant. Typically, data about research and researchers is located in database or web silos, with public and private information intermingled. Connections across departments, programs, or colleges at a university can be very hard to discern and even harder to visualize, analyze, and quantify. VIVO makes both discovery and analysis of cross-cutting information within an institution relatively simple, and this has already led a number of institutions to commit to local adoptions of VIVO.

The benefits of running VIVO at multiple cooperating institutions are even greater. VIVO and the VIVO ontology benefit significantly from the "network effect." The more institutions adopt VIVO and make their data available for interchange, the more valuable installing and using VIVO becomes. Discovering connections within an institution can be hard, but discovering, analyzing, and maintaining them across multiple institutions has previously been extraordinarily difficult. VIVO provides a way to standardize and coordinate researcher and research information across small or large groups of institutions.

VIVO is already demonstrating its value in multi-institutional settings. As was mentioned in Section 8.2.2, it has been recommended as the interchange standard by the Clinical and Translational

[163]http://libraries.mit.edu/sites/scholarly/mit-open-access/open-access-at-mit/mit-open-access-policy/
[164]http://osc.hul.harvard.edu/policies
[165]http://www.ucl.ac.uk/media/library/OpenAccess
[166]https://urresearch.rochester.edu/viewResearcherBrowse.action

Science Awards (CTSA) Consortium Executive and Steering Committee for CTSA institutions. It is being adopted at the statewide level in Colorado, and potentially by other states. The USDA has adopted VIVO for its internal researchers, and the head of the National Agricultural Library has proposed that VIVO be the mechanism for coordinating research and researchers between the USDA and the land-grant universities in all 50 states.

Typically, each consortium or regional community will coordinate the exchange of VIVO data among themselves, possibly building a cross-community search as in vivosearch.org. For broader accessibility to public VIVO data, the VIVO community could manage a registry of institutions that make available VIVO linked open data. This would allow governmental, commercial, or non-profit services to easily build on the VIVO data to support their own communities or unique value-added services.

A remaining challenge for the VIVO community is finding a home for researchers whose own institution has not adopted VIVO. We are working with both the ORCID project and the American Psychological Association's PublishTrust project[167] to find a place where those researchers can maintain their profile information. We are also looking into developments for VIVO that would better support virtual organizations or small cross-institutional research groups, with profile information coming from sources other than institutional databases.

8.5   CREATING AN OPEN-SOURCE COMMUNITY

From 2003–2009, the development of the VIVO application and ontology was primarily done at Cornell University. Since 2009, the project has expanded to include developers at Indiana University and the University of Florida. With the end of major NIH grant funding, the VIVO project is looking to move to an open-source community development model, drawing on contributors from around the world. To do this, we are working to identify an independent home for VIVO development that would not be tied to a single institution, or even a small group. Other projects have successfully made this transition, and we are seeking to work with an organization that has already done this for one or more projects.

As described in Chapter 6, a robust open-source community is growing around VIVO on SourceForge[168] and other sites, a community which supports project development efforts as well as ontology, implementation, and adoption activities. Each of these areas has regularly scheduled calls and a listserv for communication. Notably, a number of project collaborators have developed applications which enhance use of VIVO,[169] or which leverage the richly structured VIVO data,[170, 171] and have generously contributed the code to the community.

As the VIVO community grows, it is critical that new development and the evolution of VIVO be guided by the broad needs of the community. In the near term, existing VIVO project

[167] http://www.publishtrust.org
[168] http://vivo.sourceforge.net/
[169] http://sourceforge.net/p/vivo/tools/home/Home/
[170] http://vivosearchlight.org/
[171] http://xcite.hackerceo.org/VIVOviz/visualization.html

members expect to provide significant technical leadership and development expertise, but we also plan to increase the pool of developers and code committers well beyond the current institutions. To succeed, we will need formalized structures for managing tasks and releases, prioritizing development, and creating community support structures.

An independent organization would also have the potential to supply a legal and financial structure to provide a long-term home for the VIVO community. Sustainability of the technology and community is a major challenge for VIVO, and the transition from a grant-funded activity to a self-sustaining one is always a challenge. The level of adoption and interest, and the resources already being committed by a large number of institutions to their own local VIVO implementations, seem likely to result in a successful transition.

8.6   A STANDARD FOR EXCHANGING INFORMATION ABOUT RESEARCHERS

The VIVO ontology expresses highly useful and relevant information about researchers and research in a clear, structured, and extensible format. Moreover, the expression of this information as RDF triples—simple statements about researchers and their organizational and scholarly contexts—makes it very easy to exchange and add to these statements without the overhead and complexity of traditional metadata schemas. This has made the VIVO ontology an ideal mechanism for the standard interchange of researcher profile information.

Many institutions, consortia, and regional groups are already evaluating or using the VIVO ontology as their interchange standard, particularly in the area of biomedical research, where most of the major researcher-profiling systems (including VIVO, Harvard Profiles, and Loki) have adopted the VIVO ontology for exchanging profile information.

The next step is to make use of the VIVO ontology as a data- interchange standard at the national and international level. To that end, the VIVO team has been engaging with a number of different efforts and organizations.

- We are working with two related U.S. government efforts focused on researcher profiles: STAR METRICS[172] and SciENCV.[173]

- We are assisting the USDA in its adoption of VIVO [174] for its own researchers, and are in discussions with the EPA and other agencies about potential VIVO adoption.

- We have formed a partnership with euroCRIS[175] to converge the semantics of the VIVO ontology and the CERIF European Union recommended standard.

[172]https://www.starmetrics.nih.gov/
[173]http://rbm.nih.gov/profile_project.htm
[174]http://www.usda.gov/wps/portal/usda/usdahome?contentidonly=true&contentid=2010/10/0507.xml
[175]http://vivoweb.org/blog/2011/11/joint-statement-eurocris-and-vivo-project

- We have formed a partnership with CASRAI[176] and are working with them to develop a common, ontology-based approach to sharing information about research.[177]

The VIVO system and ontology have a number of significant strengths in addressing the challenges of supporting national or international interchange of researcher profiles. In particular, the information from an institutional VIVO is authoritative, typically drawing on the databases of record at the institution. It also provides a single, locally managed source for this information. Currently, researchers may need to maintain information in many different places. Once the researcher verifies the information in a local VIVO instance, that can then become the source for populating information in a wide variety of external systems.

One particular challenge is the proliferation of commercial and governmental systems that support the analysis of a researcher's publication and research record, often creating "impact factors" or other measures of researcher reputation. This challenge also represents an opportunity for VIVO: if these systems can draw on the authoritative information in a researcher's institutional VIVO profile, then they are much more likely to present a more accurate, transparent picture of scholarship and its resulting impact on the individual, group, organization, and even global level.

One final advantage to using VIVO-based data about researchers lies in the nature of linked open data. Many governments are now making significant amounts of governmental and research information available as linked open data (e.g., data.gov, data.uk.gov). There is now an opportunity to link researchers and their entire academic context into these developing datasets. One of the major strengths of linked open data is its ability to support analysis and visualization of information across very diverse kinds of data. VIVO allows authoritative data about researchers to be incorporated into these new analytic tools.

8.7   SUMMARY: VIVO'S CHALLENGES AND OPPORTUNITIES

As VIVO looks to the future and seeks to grow both its capabilities and its community it faces a number of challenges and opportunities. First, the **challenges** are presented.

Sustainability. For any project, making the shift from grant funding to sustainable development is a major challenge. While VIVO has a large number of institutions in the early stage of adoption and committing significant resources to their VIVO implementations, there is not yet a significant group of "long-term" adopters. VIVO must seek to build, nurture, and sustain both a development community and a community of adopters.

Researcher Acceptance. If individual researchers are going to put the effort into ensuring their information is correct, and if they are going to take advantage of the system for discovery and collaboration, then they need to see its value to themselves and their community.

---

[176]http://casrai.org
[177]http://vivoweb.org/blog/2012/04/vivo-joins-casrai-advancing-research-interoperability

Perceived Complexity. Compared to traditional relational database approaches, Semantic Web tools are seen as new, complex, and untried. Issues around both representation and scalability are still active areas of research and development.

Balancing these challenges are some significant **strengths and opportunities**:

Openness. The VIVO software, ontology, and data are all designed to be open—easy to use, extend, and exchange. By living in the linked open data world, VIVO information can easily be collected, related to other linked open data, and shared. By only providing public information about researchers and research, VIVO avoids significant concerns about privacy and the mechanisms for authentication and authorization that come with protecting privacy and confidential information.

Flexibility. A major advantage of the Semantic Web approach is both its flexibility and its extensibility. Interlinking new individuals, vocabularies, or external data is very simple. Extending the ontology, either for local use or for a new community, is also very straightforward. The VIVO system is designed so that the display, editing, and other capabilities extend with the ontology.

Uptake. A large number of institutions and organizations are on the path to adopt or integrate with VIVO. These include institutions of higher education from around the world, government agencies, standards groups, other researcher-profiling systems, professional societies, and publishers. VIVO is seen as a straightforward and practical solution to a common problem: creating standard descriptions of researchers and their academic context that everyone can share and reuse.

There is still a great deal of work to do before VIVO can truly fulfill its promise, but we are well down the road toward an open, flexible, and powerful solution to promoting researcher discovery, collaboration, and analysis—a solution that will benefit the entire international research community.

ACKNOWLEDGMENTS

The content of this chapter represents the work of many people who have contributed to VIVO over the years. The authors would like to thank all those whose comments, criticisms, suggestions, and direct contributions have served to advance both VIVO and the future of open data about research and researchers on the web. This work was funded in part by NIH award U24RR029822.

APPENDIX A

VIVO Ontology Classes, Object Properties, and Data Type Properties

6/30/12
VIVO 1.4.1

VIVO Ontology Classes, Object Properties, and Datatype Properties

Prefix	Ontology name	Ontology namespace	Ontology website
vivo	VIVO core ontology	http://vivoweb.org/ontology/core#	http://vivoweb.org/ontology/download
bibo	Bibliographic Ontology	http://purl.org/ontology/bibo/	http://bibliontology.com/
ero	eagle-i ontology	http://purl.obolibrary.org/obo/	http://code.google.com/p/eagle-i/
event	Event ontology	http://purl.org/NET/c4dm/event.owl#	http://motools.sourceforge.net/event/event.html
foaf	Friend of a Friend (FOAF)	http://xmlns.com/foaf/0.1/	http://xmlns.com/foaf/spec/
geo	Geopolitical ontology	http://aims.fao.org/aos/geopolitical.owl#	http://aims.fao.org/geopolitical.owl
pvs	Provenance support	http://vivoweb.org/ontology/provenance-support#	http://vivoweb.org/ontology/download
scires	Scientific research	http://vivoweb.org/ontology/scientific-research#	http://vivoweb.org/ontology/download
skos	Simple Knowledge Organization System	http://www.w3.org/2004/02/skos/core#	http://www.w3.org/2004/02/skos/

Prefix	Class	Prefix	Object Property (bold indicates explicit domain; otherwise a restriction)	Range Class	Prefix	Datatype Property (bold indicates explicit domain; otherwise a restriction)	Range Datatype (if typed)
vivo	AcademicDegree	vivo	**degreeOfferedBy**	(no range class specified)	vivo	abbreviation	
bibo	ThesisDegree	vivo	**degreeOutcomeOf**	vivo:EducationalTraining			
vivo	Address	vivo	hasGeographicLocation	vivo:GeographicLocation	vivo	**address1**	
		vivo	**mailingAddressFor**	(no range class specified)	vivo	**address2**	
					vivo	**address3**	
					vivo	addressCity	
					vivo	addressCountry	
					vivo	**addressPostalCode**	
vivo	USPostalAddress				vivo	**addressState**	
foaf	Agent	vivo	**assigneeFor**	bibo:Patent	vivo	email	
		vivo	hasAttendeeRole	vivo:AttendeeRole	vivo	**faxNumber**	
		vivo	**awardOrHonor**	vivo:AwardReceipt	vivo	freetextKeyword	
		vivo	**hasClinicalRole**	vivo:ClinicalRole	vivo	**overview**	
		vivo	**hasCollaborator**	foaf:Agent	vivo	phoneNumber	
		vivo	featuredIn	vivo:InformationResource	vivo	primaryEmail	
		vivo	**hasRole**	vivo:Role	vivo	**primaryPhoneNumber**	
		vivo	**hasLeaderRole**	vivo:LeaderRole			
		event	**isAgentIn**	event:Event			
		vivo	mailingAddress	vivo:Address			
		vivo	**hasMemberRole**	vivo:MemberRole			
		vivo	**hasOrganizerRole**	vivo:OrganizerRole			
		vivo	**hasOutreachProviderRole**	vivo:OutreachProviderRole			
		vivo	**hasPresenterRole**	vivo:PresenterRole			
		vivo	authorInAuthorship	vivo:Authorship			
		vivo	webpage	vivo:URLLink			
foaf	Group	vivo	contributingRole	vivo:Role			
vivo	Committee	vivo	hasCurrentMember	foaf:Person			
vivo	Team	vivo	hasCurrentMember	foaf:Person			
foaf	Organization	vivo	**affiliatedOrganization**	foaf:Organization	vivo	abbreviation	
		vivo	**awardConferred**	vivo:AwardReceipt			
		vivo	**awardsGrant**	vivo:Grant			
		vivo	contributingRole	vivo:Role			

Prefix	Name	Range
vivo	currentlyHeadedBy	foaf:Person
vivo	dateTimeInterval	vivo:DateTimeInterval
vivo	governingAuthorityFor	(no range class specified)
vivo	hasCurrentMember	foaf:Agent
vivo	hasEquipment	vivo:Equipment
vivo	hasGeographicLocation	vivo:GeographicLocation
vivo	hasPredecessorOrganization	foaf:Organization
vivo	hasSubOrganization	foaf:Organization
vivo	hasSuccessorOrganization	foaf:Organization
vivo	offersCourse	vivo:Course
vivo	organizationForPosition	vivo:Position
vivo	organizationForTraining	vivo:EducationalTraining
vivo	publisherOf	(no range class specified)
vivo	sponsors	vivo:Award
vivo	subOrganizationWithin	foaf:Organization
vivo	Association	
vivo	Center	
vivo	ClinicalOrganization	
vivo	College	
vivo	offersDegree	vivo:AcademicDegree
vivo	Company	
vivo	administers	vivo:Grant
vivo	PrivateCompany	
vivo	Consortium	
vivo	Department	
vivo	AcademicDepartment	
vivo	administers	vivo:Grant
vivo	offersDegree	vivo:AcademicDegree
vivo	Division	
vivo	ExtensionUnit	
vivo	administers	vivo:Grant
vivo	Foundation	
vivo	FundingOrganization	
vivo	providesFundingThrough	vivo:FundingOrganization
vivo	GovernmentAgency	
vivo	Hospital	
vivo	Institute	
vivo	Laboratory	
vivo	ResearchLaboratory (deprecated)	
vivo	CoreLaboratory	
vivo	Library	
vivo	administers	vivo:Grant
vivo	Museum	
vivo	Program	
vivo	Publisher	
vivo	ResearchOrganization	
vivo	School	
vivo	StudentOrganization	
vivo	University	
vivo	offersDegree	vivo:AcademicDegree
foaf	Person	
vivo	adviseeIn	vivo:AdvisingRelationship
vivo	advisorIn	vivo:AdvisingRelationship
vivo	awardOrHonor	vivo:AwardReceipt
vivo	currentlyHeadOf	foaf:Organization
vivo	domesticGeographicFocus	vivo:GeographicRegion
vivo	editorOf	bibo:Document
vivo	educationalTraining	vivo:EducationalTraining
vivo	eRACommonsId	
foaf	firstName	
foaf	lastName	
vivo	middleName	
vivo	orcidId	
vivo	outreachOverview	
vivo	overview	

Class		Object Property	Range	Datatype Properties
vivo FacultyMember		vivo eligibleFor	vivo:Credential	vivo preferredTitle
vivo FacultyMemberEmeritus		vivo geographicFocus	vivo:GeographicRegion	bibo prefixName
vivo ProfessorEmeritus		vivo hasCo-PrincipalInvestigatorRole	vivo:Co-PrincipalInvestigatorRole	vivo researchId
vivo Librarian		vivo hasCredential	vivo:IssuedCredential	vivo researchOverview
vivo LibrarianEmeritus		vivo hasEditorRole	vivo:EditorRole	vivo scopusId
vivo Non-Academic		vivo hasInvestigatorRole	vivo:InvestigatorRole	bibo suffixName
vivo Non-FacultyAcademic		vivo hasPrincipalInvestigatorRole	vivo:PrincipalInvestigatorRole	vivo teachingOverview
vivo Postdoc		vivo hasResearchArea	owl:Thing	
pvs PersonAsListed		vivo hasReviewerRole	vivo:ReviewerRole	
vivo Student		vivo hasServiceProviderRole	vivo:ServiceProviderRole	
vivo GraduateStudent		vivo hasTeacherRole	vivo:TeacherRole	
vivo UndergraduateStudent		vivo internationalGeographicFocus	vivo:GeographicRegion	
		vivo personInPosition	vivo:Position	
		pvs listedAuthorFor	vivo:Authorship	
vivo Agreement		vivo administeredBy	foaf:Organization	
vivo Contract		vivo contributingRole	vivo:Role	
vivo Grant		vivo dateTimeInterval	vivo:DateTimeInterval	bibo abstract
		vivo geographicFocus	vivo:GeographicRegion	vivo grantDirectCosts
		vivo hasSubjectArea	owl:Thing	vivo localAwardId
		vivo fundingVehicleFor	(no range class specified)	vivo sponsorAwardId
		vivo grantAwardedBy	foaf:Organization	vivo totalAwardAmount
		vivo grantSubcontractedThrough	foaf:Organization	
		vivo hasSubGrant	vivo:Grant	
		vivo subGrantOf	vivo:Grant	
		vivo supportedInformationResource	vivo:InformationResource	
		vivo webpage	vivo:URLLink	
geo area		geo isPredecessorOf	geo:area	geo (36 datatype properties)
		geo isSuccessofOf	geo:area	
geo group		geo hasMember	geo:territory	geo (36 datatype properties)
geo economic_region				
geo organization				
geo special_group				
geo geographical_region				
geo territory		geo hasBorderWith	geo:territory	geo (89 datatype properties)
geo disputed		geo isInGroup	geo:group	
geo non_self_governing				
geo other		geo isAdministeredBy	geo:self_governing	

geo self_governing

Class		Object property		Range	Data property	
vivo	Award	vivo	awardConferredBy	foaf:Organization	vivo	description
		vivo	**sponsoredBy**	foaf:Organization		
		vivo	**receipt**	vivo:AwardReceipt		
vivo	AwardReceipt	vivo	awardConferredBy	foaf:Organization	vivo	description
		vivo	awardOrHonorFor	foaf:Agent		
		vivo	dateTimeInterval	vivo:DateTimeInterval		
		vivo	dateTimeValue	vivo:DateTimeValue		
		vivo	receiptOf	vivo:Award		
	ERO_0000020 (biological specimen)					
skos	Concept	skos	broader	skos:Concept		
		skos	narrower	skos:Concept		
		skos	related	skos:Concept		
vivo	Credential	vivo	hasGoverningAuthority	foaf:Organization		
		vivo	hasSubjectArea	owl:Thing		
		vivo	issuedCredential	vivo:IssuedCredential		
		vivo	validIn	vivo:GeographicLocation		
vivo	Certificate					
vivo	License					
vivo	DateTimeInterval	vivo	end	vivo:DateTimeValue		
		vivo	start	vivo:DateTimeValue		
vivo	AcademicTerm					
vivo	AcademicYear					
vivo	DateTimeValue	vivo	dateTimePrecision	vivo:DateTimeValuePrecision	vivo	**dateTime** datetime
vivo	DateTimeValuePrecision					
bibo	DocumentStatus					
vivo	EducationalTraining	vivo	**contributingAdvisingRelationship**	vivo:AdvisingRelationship	vivo	**departmentOrSchool**
		vivo	dateTimeInterval	vivo:DateTimeInterval	vivo	**majorField**
		vivo	**degreeEarned**	vivo:AcademicDegree	vivo	**supplementalInformation**
		vivo	**trainingAtOrganization**	foaf:Organization		
		vivo	**educationalTrainingOf**	foaf:Person		
vivo	Internship					
vivo	MedicalResidency					
vivo	PostdoctoralTraining					
vivo	Equipment	vivo	**equipmentFor**	foaf:Organization	vivo	freetextKeyword
		vivo	**equipmentInFacility**	vivo:Facility		
		vivo	webpage	vivo:URLLink		
		ero	ERO_0000460 (has documentation)	bibo:Document		
ero	ERO_0000004 (instrument)					
event	Event	event	**agent**	foaf:Agent	vivo	contactInformation
		vivo	dateTimeInterval	vivo:DateTimeInterval	vivo	description
		vivo	domesticGeographicFocus	vivo:GeographicRegion		
		vivo	**eventWithin**	event:Event		
		vivo	geographicFocus	vivo:GeographicRegion		
		vivo	hasGeographicLocation	vivo:GeographicLocation		
		vivo	hasSubjectArea	owl:Thing		
		vivo	**includesEvent**	event:Event		
		vivo	**inEventSeries**	vivo:EventSeries		
		vivo	internationalGeographicFocus	vivo:GeographicRegion		
		vivo	realizedRole ('participant')	vivo:Role		

Class	Object Property	Range	Data Type Property
vivo Competition	bibo presents ('related documents')	bibo:Document	
	vivo webpage	vivo:URLLink	
bibo Conference	vivo hasProceedings	bibo:Proceedings	vivo abbreviation
vivo Course	vivo hasPrerequisite	foaf:Course	vivo courseCredits
	vivo courseOfferedBy	foaf:Organization	
	vivo prerequisiteFor	vivo:Course	
vivo Exhibit			
bibo Hearing			
bibo Interview			
vivo Meeting			
bibo Performance	bibo performer	foaf:Agent	
vivo Presentation			
vivo InvitedTalk			
bibo Workshop			
vivo EventSeries	vivo dateTimeInterval	vivo:DateTimeInterval	vivo contactInformation
	vivo domesticGeographicFocus	vivo:GeographicRegion	vivo description
	vivo geographicFocus	vivo:GeographicRegion	
	vivo hasSubjectArea	owl:Thing	
	vivo internationalGeographicFocus	vivo:GeographicRegion	
	vivo realizedRole ('participant')	vivo:Role	
	vivo seriesForEvent	vivo:EventSeries	
	vivo webpage	vivo:URLLink	
vivo ConferenceSeries			
vivo SeminarSeries			
vivo WorkshopSeries			
vivo InformationResource	vivo dateTimeValue	vivo:DateTimeValue	vivo freetextKeywords
	vivo domesticGeographicFocus	vivo:GeographicRegion	
	bibo editor	foaf:Person	
	vivo features	(no range class specified)	
	vivo geographicFocus	vivo:GeographicRegion	
	vivo hasSubjectArea	owl:Thing	
	vivo informationResourceInAuthorship	vivo:Authorship	
	vivo informationResourceProductOf	(no range class specified)	
	vivo informationResourceSupportedBy	(no range class specified)	
	vivo internationalGeographicFocus	vivo:GeographicRegion	
	bibo translator	(no range class specified)	
	vivo webpage	vivo:URLLink	
bibo Collection			
bibo Periodical	vivo publisher	foaf:Organization	bibo oclcnum
			bibo eissn
			bibo issn
bibo Code			
bibo CourtReporter			
bibo Journal			vivo abbreviation
bibo Magazine			
vivo Newsletter			
bibo Newspaper			
bibo Series			
bibo Website	vivo publisher	foaf:Organization	
vivo Blog			

vivo Dataset
bibo Document

bibo	citedBy	bibo:Document
bibo	cites	bibo:Document
scires	documentationFor	(no range class specified)
vivo	editor	foaf:Person
vivo	hasPart	bibo:DocumentPart
bibo	presentedAt	event:Event
vivo	hasPublicationVenue	bibo:Collection
bibo	reproducedIn	bibo:Document
vivo	reproduces	bibo:Document
bibo	status	bibo:DocumentStatus
bibo	transcriptOf	bibo:Document

bibo	abstract
bibo	edition
bibo	locator
bibo	number
bibo	numPages
bibo	pageEnd
bibo	pageStart
bibo	pmid
bibo	section
bibo	volume

bibo Article

bibo	issue
bibo	doi
vivo	nihmsid
vivo	pmcid

bibo AcademicArticle — vivo publishedIn → bibo:Journal
vivo BlogPosting — vivo publishedIn → vivo:Blog
vivo ConferencePaper — bibo presentedAt → event:Event — bibo eanucc13
vivo EditorialArticle
vivo Review

bibo AudioDocument — bibo distributor → foaf:Organization
bibo AudioVisualDocument — bibo distributor → foaf:Organization
bibo Film
vivo Video

bibo Book — vivo publisher → foaf:Organization

bibo	isbn10
bibo	isbn13
bibo	lccn
vivo	placeOfPublication

bibo Proceedings — vivo proceedingsOf → bibo:Conference
vivo CaseStudy
vivo Catalog — vivo publisher → foaf:Organization — vivo placeOfPublication
bibo CollectedDocument — vivo publisher → foaf:Organization
vivo Database
bibo EditedBook — vivo placeOfPublication
bibo Issue
vivo ConferencePoster
bibo DocumentPart — vivo publisher → foaf:Organization — vivo placeOfPublication
bibo BookSection — vivo part of → bibo:Document
bibo Chapter — bibo chapter
bibo Excerpt
bibo Quote
bibo Slide
bibo Image
bibo Map
bibo LegalDocument
bibo LegalCaseDocument — bibo court → foaf:Organization
bibo Brief
bibo Decision — bibo affirmedBy → bibo:Decision

Class	Object Properties	Range	Data Type Properties
bibo Legislation			
bibo Bill			
bibo Statute			
bibo Manual			
bibo Manuscript			
vivo NewsRelease			
bibo Note	bibo annotates	(no range class specified)	
bibo Patent	vivo assignee	foaf:Agent	vivo cclCode
	vivo dateFiled	vivo:DateTimeValue	vivo iclCode
	vivo dateIssued	vivo:DateTimeValue	vivo **patentNumber**
	vivo publisher	foaf:Organization	
ero OBI_0000272 (protocol)			
bibo ReferenceSource	bibo distributor	foaf:Organization	
bibo Report			vivo reportId
vivo ResearchProposal			
vivo Score			
vivo Screenplay			
bibo Slideshow			
vivo Speech			
bibo Standard			
bibo Thesis	vivo relatedDegree	bibo:ThesisDegree	vivo abbreviation
			vivo placeOfPublication
vivo Translation			
bibo Webpage			vivo placeOfPublication
vivo WorkingPaper			
foaf Image (deprecated in favor of bibo:Image)			vivo description
vivo Software			vivo description
vivo IssuedCredential	vivo credentialOf	foaf:Person	
	vivo dateIssued	vivo:DateTimeValue	
	vivo expirationDate	vivo:DateTimeValue	
	vivo hasSubjectArea	owl:Thing	
	vivo issuanceOf	vivo:Credential	
	vivo validIn	vivo:GeographicLocation	
vivo Certification			
vivo Licensure			vivo **licenseNumber**
vivo Location			
vivo GeographicLocation	vivo geographicallyContains	vivo:GeographicLocation	
	vivo geographicallyWithin	vivo:GeographicLocation	
	vivo geographicLocationOf	(no range class specified)	
vivo Campus			
vivo Facility	vivo facilityFor	(no range class specified)	
	vivo locationOfEquipment	vivo:Equipment	
vivo Building	vivo hasRoom	vivo:Room	
vivo Room	vivo roomWithingBuilding	vivo:Building	vivo **seatingCapacity** int
vivo GeographicRegion	vivo domesticGeographicFocusOf	(no range class specified)	
	vivo geographicFocusOf	(no range class specified)	
	vivo internationalGeographicFocusOf	(no range class specified)	
vivo GeopoliticalEntity	vivo geographicallyWithin	vivo:GeographicLocation	geo (89 datatype properties)
vivo Country	geo hasBorderWith	geo:territory	
= geo self_governing			

vivo County
vivo PopulatedPlace
vivo StateOrProvince

geo isInGroup — geo:group
geo isPredecessorOf — geo:area
geo isSuccessorOf — geo:area
vivo geographicallyWithin — vivo:StateOrProvince

geo (89 datatype properties)

geo territory

geo hasBorderWith — geo:territory
geo isInGroup — geo:territory
geo isPredecessorOf — geo:area
geo isSuccessorOf — geo:area

geo (89 datatype properties)

geo disputed
geo non_self_governing
geo other
geo self_governing

geo isAdministeredBy — geo:self_governing

vivo SubnationalRegion
vivo TransnationalRegion
vivo Continent
geo geographical_region

geo hasMember — geo:group
geo isPredecessorOf — geo:area
geo isSuccessorOf — geo:area

geo (36 datatype properties)

ero OBI_0100026 (organism)
vivo Position
vivo FacultyAdministrativePosition
vivo FacultyPosition
vivo LibrarianPosition
vivo NonAcademicPosition
vivo NonFacultyAcademicPosition
vivo PostdoctoralPosition

vivo email
hrJobTitle
vivo rank — int

vivo Project

vivo dateTimeInterval — vivo:DateTimeInterval
vivo domesticGeographicFocus — vivo:GeographicRegion
vivo hasFundingVehicle — vivo:Grant
vivo geographicFocus — vivo:GeographicRegion
vivo informationProduct ('produces') — vivo:InformationResource
vivo internationalGeographicFocus — vivo:GeographicRegion
vivo realizedRole ('participant') — vivo:Role
vivo webpage — vivo:URLLink

vivo contactInformation
vivo description

ero ERO_0000014 (research project)
ero ERO_0000015 (human study)
ero ERO_0000016 (clinical trial)

scires Phase0ClinicalTrial
scires Phase1ClinicalTrial
scires Phase2ClinicalTrial
scires Phase3ClinicalTrial
scires Phase4ClinicalTrial

scires irbNumber
scires nctId
scires studyPopulationCount

ero ERO_0000006 (reagent)
vivo Relationship
vivo AdvisingRelationship

foaf:Person
vivo:EducationalTraining

vivo advisee
vivo advisingContributionTo

Class		Object Property		Range	Data Type Property		Type
vivo	FacultyMentoringRelationship						
vivo	GraduateAdvisingRelationship						
vivo	PostdocOrFellowAdvisingRelationship						
vivo	UndergraduateAdvisingRelationship	vivo	advisor	foaf:Person			
		vivo	dateTimeInterval	vivo:DateTimeInterval			
		vivo	degreeCandidacy	vivo:AcademicDegree			
		vivo	hasSubjectArea	owl:Thing			
vivo	Authorship	pvs	authorAsListed	pvs:PersonAsListed	vivo	authorRank	int
		vivo	linkedAuthor	foaf:Agent	vivo	hideFromDisplay	boolean
		vivo	linkedInformationResource	vivo:InformationResource	vivo	isCorrespondingAuthor	boolean
ero	ERO_0000295 (research opportunity)	vivo	dateTimeInterval	vivo:DateTimeInterval	vivo	contactInformation	
		vivo	hasFundingVehicle	vivo:Grant	vivo	description	
vivo	Role	vivo	associatedWithPosition	vivo:Position	vivo	description	
		vivo	roleContributesTo	(no range class specified)	vivo	hideFromDisplay	boolean
		vivo	dateTimeInterval	vivo:DateTimeInterval			
		vivo	roleRealizedIn	(no range class specified)			
		vivo	RoleOf	foaf:Agent			
vivo	AttendeeRole	vivo	attendeeRoleOf	(no range class specified)			
vivo	ClinicalRole	vivo	clinicalRoleOf	foaf:Agent			
vivo	EditorRole	vivo	editorRoleOf	foaf:Person			
vivo	LeaderRole	vivo	leaderRoleOf	foaf:Agent			
vivo	MemberRole	vivo	memberRoleOf	foaf:Agent			
vivo	OrganizerRole	vivo	organizerRoleOf	foaf:Agent			
vivo	OutreachProviderRole	vivo	outreachProviderRoleOf	foaf:Agent			
vivo	PresenterRole	vivo	presenterRoleOf	foaf:Person			
vivo	ResearcherRole	vivo	researcherRoleOf	foaf:Person			
vivo	InvestigatorRole	vivo	investigatorRoleOf	foaf:Person			
vivo	CoPrincipalInvestigatorRole	vivo	co-PrincipalInvestigatorRoleOf	foaf:Person			
vivo	PrincipalInvestigatorRole	vivo	principalInvestigatorRoleOf	foaf:Person			
vivo	ReviewerRole	vivo	reviewerRoleOf	foaf:Person			
vivo	PeerReviewerRole						
vivo	ServiceProviderRole	vivo	serviceProviderRoleOf	foaf:Person			
vivo	TeacherRole	vivo	teacherRoleOf	foaf:Person			
vivo	Service	vivo	contributingRole	vivo:Role			
		vivo	serviceProvidedBy	(no range class specified)			
ero	ERO_0000391 (access service)	ero	ERO_0000481 ('realizes protocol')	ero:Protocol			
ero	ERO_0000394 (production service)	vivo	webpage	vivo:URLLink			
ero	ERO_0000392 (storage service)	ero	ERO_0000029 ('provides access to')	(no range class specified)			
vivo	URLLink	vivo	webpageOf	(no range class specified)	vivo	linkAnchorText	
					vivo	linkURI	
					vivo	rank	int

VIVO core ontology as of version 1.4.1; for an updated version, please visit http://sourceforge.net/apps/mediawiki/vivo/index.php?title=Ontology#Ontology_Documentation

Prefix	Class / Property	Range
vivo	dateTimeInterval	DateTimeInterval
vivo	roleRealizedIn	(no range class specified)
vivo	RoleOf	Agent
vivo	AttendeeRole	
vivo	attendeeRoleOf	(no range class specified)
vivo	ClinicalRole	
vivo	clinicalRoleOf	Agent
vivo	EditorRole	
vivo	editorRoleOf	Person
vivo	LeaderRole	
vivo	leaderRoleOf	Agent
vivo	MemberRole	
vivo	memberRoleOf	Agent
vivo	OrganizerRole	
vivo	organizerRoleOf	Agent
vivo	OutreachProviderRole	
vivo	outreachProviderRoleOf	Agent
vivo	PresenterRole	
vivo	presenterRoleOf	Person
vivo	ResearcherRole	
vivo	researcherRoleOf	Person
vivo	InvestigatorRole	
vivo	investigatorRoleOf	Person
vivo	CoPrincipalInvestigatorRole	
vivo	co-PrincipalInvestigatorRoleOf	Person
vivo	PrincipalInvestigatorRole	
vivo	principalInvestigatorRoleOf	Person
vivo	ReviewerRole	
vivo	reviewerRoleOf	Person
vivo	PeerReviewerRole	
vivo	ServiceProviderRole	
vivo	serviceProviderRoleOf	Person
vivo	TeacherRole	
vivo	teacherRoleOf	Person
vivo	contributingRole	Role
vivo	Service	
vivo	serviceProvidedBy	(no range class specified)
ero	ERO_0000391 (access service)	
ero	ERO_0000394 (production service)	
ero	ERO_0000392 (storage service)	
ero	ERO_0000481 ("realizes protocol")	Protocol
vivo	webpage	URLLink
ero	ERO_0000029 ("provides access to")	(no range class specified)
vivo	URLLink	
vivo	webpageOf	(no range class specified)
vivo	linkAnchorText	
vivo	linkURI	
vivo	rank	int

VIVO core ontology as of version 1.4.1; for an updated version, please visit http://sourceforge.net/apps/mediawiki/vivo/index.php?title=Ontology#Ontology_Documentation

Authors' Biographies

PAUL J. ALBERT

Paul J. Albert is Assistant Director of Research and Digital Services at Weill Cornell Medical Library. He serves as information architect for Weill Cornell's implementation of VIVO and as a member of VIVO's national Ontology Team. Paul's interests include publication metadata and repurposing VIVO data in novel and useful ways.

CHRIS BARNES

Chris Barnes is the Associate Director of the Clinical and Translational Research Informatics Program. On the VIVO grant, he served as a team lead for development at the University of Florida focusing on internal and external integration and packaging for VIVO. His work and research at Florida is in the area of biomedical informatics.

JIM BLAKE

Jim Blake has been a member of the VIVO team at Cornell University since 2009. He is a software developer and systems engineer, with experience that extends from military flight simulators to handicapping systems for horse races. He currently describes himself as "a singer, a dancer, and a middle-tier Java jockey."

KATY BÖRNER

Katy Börner is the Victor H. Yngve Professor of Information Science at the School of Library and Information Science, Adjunct Professor at the School of Informatics and Computing, Adjunct Professor at the Department of Statistics in the College of Arts and Sciences, and Founding Director of the Cyberinfrastructure for Network Science Center (http://cns.iu.edu) at Indiana University. Her research focuses on the development of data analysis and visualization techniques for information access, understanding, and management. Within the VIVO project, she led the team at Indiana University and directed the development of social network visualizations.

BRIAN CARUSO

Brian Caruso is a software developer at Cornell University's Albert R. Mann Library and has worked on the VIVO project since 2005. He was lead of the application development team at Mann from

2009–2011. He continues to work on VIVO with interests in system architecture, generalization of editing, sharing data, and search.

CURTIS COLE

Curtis Cole is currently the Chief Information Officer at Weill Cornell Medical College, where he is responsible for the core information services that support the research, clinical, educational, and administrative functions of the college. Previously, as Chief Medical Information Officer, he led the implementation of a new electronic medical record system. He is also actively involved in the development of computer systems that support Clinical Research and Terminology Services.

MICHAEL CONLON

Michael Conlon is Associate Director and Chief Operating Officer of the University of Florida Clinical and Translational Science Institute, Director of Biomedical Informatics at the UF College of Medicine, and Principal Investigator of the NIH project "VIVO: Enabling National Networking of Scientists." His responsibilities include development of academic biomedical informatics, expansion and integration of research and clinical information resources, and strategic planning for academic health and university research. As PI of the VIVO project, Dr. Conlon leads a team of 120 investigators at 7 schools in the development, implementation, and advancement of an open-source Semantic Web application for research discovery. He earned his PhD in Statistics from the University of Florida, undergraduate degrees in Mathematics and Economics from Bucknell University, and is the author of over 150 scholarly publications and presentations. His current interests include enterprise change and organizational issues in the adoption of information technology, organization of research resources, and enterprise architecture. See vivo.ufl.edu/individual/mconlon for connections, works, and shared scholarly information.

JON CORSON-RIKERT

Jon Corson-Rikert is the head of Information Technology Services at Cornell University's Albert R. Mann Library. He was the initial VIVO developer at Cornell starting in 2003 and has served as the development lead and a member of the ontology team for the NIH project "VIVO: Enabling National Networking of Scientists." He holds a Bachelor's degree in Visual and Environmental Studies from Harvard University and worked in cartography, geographic information systems, and computer graphics before joining the Cornell University Library in 2001.

VALRIE DAVIS

Valrie Davis is the Outreach Librarian for Agricultural Sciences at the University of Florida's Marston Science Library. She served as the National Implementation Lead for the VIVO grant and as the Local Implementation Lead for VIVO at the University of Florida. Her research focuses on the challenges of increasing access to information resources within the Agricultural Sciences community. She holds a Master's degree in Library Science from Florida State University.

YING DING

Ying Ding is an Assistant Professor at the School of Library and Information Science, Indiana University. Previously, she worked as a senior researcher at the University of Innsbruck, Austria, and as a researcher at the Free University of Amsterdam, the Netherlands. She has been involved in various NIH-funded and European Union-funded Semantic Web projects. She has published 130+ papers in journals, conferences, and workshops. She is the coeditor of a book series called *Semantic Web Synthesis* from Morgan & Claypool Publishers. She is co-author of the book *Intelligent Information Integration in B2B Electronic Commerce* and is co-author of book chapters in *Spinning the Semantic Web* and *Towards the Semantic Web: Ontology-driven Knowledge Management*. Her current interest areas include social network analysis, Semantic Web, citation analysis, knowledge management, and the application of web technology.

CHUN GUO

Chun Guo is a graduate student at the School of Library and Information Science, Indiana University. She has been working on the VIVO ontology team for two years. Her current interest areas include information retrieval and Semantic Web.

KRISTI L. HOLMES

Kristi L. Holmes is a bioinformaticist at Washington University's Becker Medical Library, where she works to develop and support cross-disciplinary initiatives across a variety of subject areas and audiences. Her professional interests include educational program development and implementation; collaboration support; open science; and research impact (http://becker.wustl.edu/impact-assessment). She currently serves at the National Outreach Lead for the VIVO project as well as the Local Outreach Lead for Washington University. She holds a PhD in Biochemistry from Iowa State University.

CHIN HUA KONG

Chin Hua Kong joined the Cyberinfrastructure for Network Science Center at Indiana University in July 2010 and has led the center's software development team since May 2011. He contributed to and led development efforts for several NSF, NIH, and Bill & Melinda Gates Foundation funded projects, including Mapping Sustainability (http://mapsustain.cns.iu.edu); International Researcher Networking (IRN) at http://nrn.cns.iu.edu; Scaling Philanthropy (http://www.milliondollarlist.org); and CIShell powered tools (http://cishell.org) such as EpiC, the Science of Science (Sci2) tool, and a web services platform for the analysis and visualization of scholarly data. He has six years of research and industrial experience in distributed systems, firmware engineering, network media streaming, and enterprise search services. His main interests are software as a service, social networking, and information visualization with a particular focus on the impact of software design on social networking and the design of insightful visualizations. He holds a B.S. in Mathematics from the University of Science in Malaysia, in 2002, and an M.S. in Computer Science from Indiana University, in 2010.

DEAN B. KRAFFT

Dean B. Krafft is the Chief Technology Strategist and Director of Information Technology at the Cornell University Library, where he helps the Library to navigate the disruptive transition from physical to digital. He served as the lead on the Cornell University subcontract for the NIH project "VIVO: Enabling National Networking of Scientists." His previous research has focused on digital libraries, including serving as Principal Investigator on the National Science Digital Library Core Integration project at Cornell. Dr. Krafft holds a PhD in Computer Science from Cornell University.

MICAH LINNEMEIER

Micah Linnemeier was the software development team lead at the Cyberinfrastructure for Network Science Center at Indiana University, where he contributed to and led development efforts for several NSF- and NIH-funded projects, including Network Workbench (http://nwb.cns.iu.edu), EpiC tool (http://epic.cns.iu.edu), the Science of Science (Sci2) (http://sci2.cns.iu.edu), and VIVO International Researcher Network (http://vivoweb.org). He holds a BS in Computer Science from Indiana University and is currently pursuing an MS in Information with a specialization in Human-Computer Interaction at the University of Michigan.

BRIAN J. LOWE

Brian J. Lowe is a software developer at Cornell University's Albert R. Mann Library. He has worked on VIVO and its underlying software engine since January 2006, and, in collaboration with Brian Caruso, engineered its transition to a fully Semantic Web-based application in 2007. Brian served as the VIVO semantic development team lead from 2009–2011 and continues to guide the evolution of VIVO's semantic components. Brian contributes to the development of the VIVO ontology, and his interests include maximizing the use of ontology semantics for practical benefit in transforming and sharing data.

LESLIE MCINTOSH

Leslie McIntosh is a Research Instructor in Pathology and Immunology at the Center for Biomedical Informatics at Washington University. Leslie served as the evaluator of the national VIVO project and serves as the Implementation Lead for Washington University. Leslie holds a PhD in Public Health Epidemiology from Saint Louis University.

STELLA MITCHELL

Stella Mitchell is a member of the VIVO team at Cornell University working on semantic software development and ontology design. Her interests include the Semantic Web and its application in higher education, environmental, biomedical and clinical domains. She holds a Bachelor's degree in Physics and Philosophy from Cornell University and an MS in Computer Science from SUNY at Binghamton. Before joining Cornell University Library in 2010, she spent 20 years in IBM research and product development.

NARAYAN RAUM

Narayan Raum is the Assistant Director of the Clinical and Translational Research Informatics Program. On the VIVO grant, he served as a team lead for development at the University of Florida focusing on internal and external integration and data harvesting for VIVO. His work and research at Florida is in the area of biomedical informatics and software engineering.

NICHOLAS REJACK

Nicholas Rejack is the VIVO Data Steward at the University of Florida's Clinical and Translational Science Institute. His responsibilities include managing the local data and ontology for VIVO and other semantic projects. He joined the University of Florida in 2010. He earned his MSIS degree from the University of Texas at Austin and his BA in Linguistics from McGill University. His interests include the Semantic Web and its applications in research and higher education, biomedical ontologies, and linked open data and its creative reuse.

NICHOLAS SKAGGS

Nicholas Skaggs serves as the QA Community Coordinator at Canonical Ltd for the Ubuntu Linux community. He served on the development team for VIVO at the University of Florida, focusing on data integration, community, and packaging for the VIVO project. Today he continues as a project member and an advocate of the power of open source and community. His interests include big data, open-source development, community building, and software development and quality assurance methodologies.

VINCENT SPOSATO

Vincent Sposato is a Senior Software Engineer and Team Lead with the Academic Health Center's Enterprise Software Engineer group. His primary focus has been in the development of reproducible harvests, improvement of the harvester tool set, and support of the UF VIVO implementation.

CHINTAN TANK

Chintan Tank received an MS in Computer Science from Indiana University in 2009, where his main interests were in the fields of web mining and computer graphics. After graduating, he worked as a software developer in the Cyberinfrastructure for Network Science Center initially implementing diverse algorithms for CIShell powered tools (http://cishell.org) and later becoming one of the lead programmers for VIVO. Among others, he implemented sparklines, egocentric and temporal graph visualizations and designed a visualization architecture that makes it easy for third-party developers to add visualizations to VIVO and leverage the power of VIVO's semantic data. In September 2011, he became the User Interface Lead at the media-analytics company General Sentiment.

LIZ TOMICH

Liz Tomich is the Director of the FIS Development Team at the University of Colorado Boulder. Liz has worked for CU–Boulder since the early years of FIS in the late 1990s and has always promoted the qualities of collaboration, teamwork, and new possibilities. Liz holds a BS in Food Science & Technology from the University of California at Davis, explored epidemiology prior to joining the Office of Faculty Affairs, and has spent much of her career playing with data.

ALEX VIGGIO

Alex Viggio is the lead developer of the Faculty Information System team at the University of Colorado Boulder. Alex's career in software development spans more than two decades in roles ranging from employee to consultant to founder. A focus on the user experience and how that relates to software quality has been a common thread in industry sectors including Internet services, information technology, financial services, telecommunications, non-profits, and transportation. His professional interests include community building, collaborative software, knowledge discovery in databases, content management, open-source advocacy, and software development best practices. He holds a BS in Computer Systems from Florida Atlantic University, 1994, and is working on an MS in Information and Communication Technology for Development from the University of Colorado Boulder.

STEPHEN WILLIAMS

Stephen Williams is a Senior Software Engineer and Team Lead with the Academic Health Center's Enterprise Software Engineering. On the VIVO grant, he served initially as National Implementation IT Support at Marston Science Library before moving into a developer position at Clinical and Translational Research Informatics Program focusing on internal and external integration and packaging for VIVO. He holds a B.S. degree in Computer Science from the University of Florida.

Printed in the United States
by Baker & Taylor Publisher Services